The Impact of Building Information Modelling

Construction projects involve complex sets of relationships between parties with different professional backgrounds trying to achieve very complex goals. Under these difficult circumstances, the quality of information on which projects are based should be of the highest possible standard. The line-based, two-dimensional drawings on which conventional construction is based render this all but impossible. This is the source of some major shortcomings in the construction industry, and this book focuses on the two most fundamental of these: the failure to deliver projects predictably, to the required quality, on time and within budget; and the failure of most firms in the industry to make a survivable level of profit. By transforming the quality of information used in building, Building Information Modelling (BIM) promises to transform construction more or less completely.

After describing and explaining these problems, the way in which BIM promises to provide solutions is examined in detail. A discussion of the theory and practice of BIM is also provided, followed by a review of various recent surveys of BIM usage in the US, UK and selected European economies. The way in which other industries, including retail and manufacturing, have been transformed by information are explored and compared with current developments in the deployment of BIM in construction. Five case studies from the UK show how BIM is being implemented, and the effects it is having on architects and contractors.

This book is perfect for any construction professional interested in improving the efficiency of their business, as well as undergraduate and postgraduate students wishing to understand the importance of BIM.

Ray Crotty worked in a variety of management control roles on North Sea projects with Bechtel, Phillips Petroleum and Shell, before going on to spend 10 years with Bovis. He devised and implemented the extranet-based document management and communications systems – the earliest known project collaboration system – used on the Bluewater project in Kent, England. He founded C3 Systems Ltd in 1999 to develop the Bluewater ideas and to generalise their use in the industry. He was a founding member of the UK chapter of BuildingSMART (formerly IAI).

The Impact of Building Information Modelling
Transforming construction

Ray Crotty

Spon Press
an imprint of Taylor & Francis
LONDON AND NEW YORK

First published 2012
by SPON Press
2 Park Square, Milton Park, Abingdon, Oxon OX14 4RN

Simultaneously published in the USA and Canada
by Routledge
711 Third Avenue, New York, NY 10017

Routledge is an imprint of the Taylor & Francis Group, an informa business

© 2012 Ray Crotty

The right of Ray Crotty to be identified as the author of this work has been asserted in accordance with sections 77 and 78 of the Copyright, Designs and Patents Act.

All rights reserved. No part of this book may be reprinted or reproduced or utilised in any form or by any electronic, mechanical, or other means, now known or hereafter invented, including photocopying and recording, or in any information storage or retrieval system, without permission in writing from the publishers.

Trademark notice: Product or corporate names may be trademarks or registered trademarks, and are used only for identification and explanation without intent to infringe.

British Library Cataloguing in Publication Data
A catalogue record for this book is available from the British Library

Library of Congress Cataloging in Publication Data
Crotty, Ray.
The impact of building information modelling : transforming construction / Ray Crotty.
    p. cm.
  1. Building information modeling. 2. Building–Data processing.
  3. Construction industry–Information resources management. I. Title.
TH438.13.C76 2011
690.01´1–dc22   2011009575

ISBN: 978-0-415-60167-2 (hbk)
ISBN: 978-0-203-83601-9 (ebk)

Typeset in Goudy
HWA Text and Data Management, London

 Printed and bound in Great Britain by the MPG Books Group

For Martha

Contents

Illustrations x
Foreword xii
Acknowledgements xiv
Picture credits and sources xv

1 Introduction 1
1.0 *Introduction: problems with drawings* 1
1.1 *BIM modelling systems* 3
1.2 *BIM and standards* 4
1.3 *BIM in action: Ikea kitchens* 6
1.4 *The advantages of model-based design* 7
1.5 *The bigger picture* 10

2 UK industry background 14
2.0 *Introduction* 14
2.1 *Key background features of UK construction* 14
2.2 *Strategic challenges* 23
2.3 *Solutions* 25

3 The problem 30
3.0 *Introduction* 30
3.1 *General features of drawing-based design information* 30
3.2 *The impact of poor information on design processes* 32
3.3 *The impact of poor information on the procurement process* 36
3.4 *The impact of poor information on construction management* 39

4 The solution 44
4.0 *Introduction* 44
4.1 *General features of BIM-based design* 44
4.2 *The advantages of BIM-based design information* 46

    4.3 *The advantages of BIM in contract procurement* 50
    4.4 *The advantages of BIM in construction management* 50
    4.5 *Production management in construction* 54
    4.6 *Conclusion* 57

5 The origins of BIM in computer-aided design   58
    5.0 *Introduction* 58
    5.1 *Terms clarified* 58
    5.2 *CAD application areas – key challenges* 59
    5.3 *A brief history of computer-aided design* 62

6 Building Information Modelling   81
    6.0 *Introduction* 81
    6.1 *BIM authoring tools – characteristics of BIM systems* 82
    6.2 *Construction project software* 88
    6.3 *Information management on BIM projects* 91
    6.4 *Sources of BIM implementation guidance* 106
    6.5 *Conclusion* 107

7 BIM – the current state of play   109
    7.0 *Introduction* 109
    7.1 *Surveys* 109
    7.2 *Case studies: introduction* 115
    7.3 *Case study: Frank Gehry's architecture* 116
    7.4 *Case study: Ryder Architecture* 120
    7.5 *Case study: Ramboll* 124
    7.6 *Case study: Team Homes Limited, Parmiter Street development* 128
    7.7 *Case study: Llanelli Scarlets Rugby Stadium – Parc y Scarlets* 132
    7.8 *Conclusions* 134

8 IT usage in construction and other industries   136
    8.0 *Introduction* 136
    8.1 *The digital revolution – changing the nature of work* 137
    8.2 *The diffusion of innovations* 138
    8.3 *General patterns of IT adoption in industry* 140
    8.4 *Major industries transformed by information technology* 146
    8.5 *Social consequences* 160
    8.6 *The pattern of IT adoption by construction firms* 166

9 Looking forward: building with perfect information   **177**
   9.0 Introduction 177
   9.1 Future construction 181
   9.2 Considerations 188
   9.3 The global picture 198

Index   *205*

Illustrations

Tables

6.1	BIM authoring applications	89
6.2	Content of key project information flows	97
6.3	Design stages: indicative level of detail	105
7.1	McGraw-Hill reports: adoption intensity	111
8.1	Information needed for construction project management	170

Figures

2.1	Construction output and GDP 1956–2006	19
2.2	UK construction – top firms, by contracting turnover	21
2.3	UK project predictability, 2000–2010	24
2.4	Major contractors' contracting margins 1987–2009	25
3.1	Information and communications in construction	33
4.1	MacLeamy curve	46
4.2	The production management approach	56
5.1	A parametric object	73
6.1	Information generated on a construction project	82
6.2	BIM – a three-layer approach	83
6.3	Construction project software map	90
6.4	Key project information flows	96
6.5	Key design deliverables by RIBA Plan of Work stage	104
6.6	Bew–Richards BIM diagram	107
7.1	Distribution of 2D and 3D CAD usage, USA, 2010	114
7.2	Weisman Museum, Minneapolis	117
7.3	Gehry's Fish, Barcelona Olympic Village	118
7.4	Guggenheim Museum, Bilbao	119
7.5	Guggenheim Museum, Bilbao	119
7.6	Victoria Hall student accommodation – 3D BIM model	122
7.7	Manchester Central Library – visualisation derived from BIM model	123
7.8	Grimsby University Centre – multi-disciplinary model	124
7.9	Norwich Open Academy	125

7.10	Wakefield Waterfront – before	127
7.11	Wakefield Waterfront – after	127
7.12	The Hepworth Gallery, Wakefield	127
7.13	Team Homes, Parmiter Street development	129
7.14	Parmiter Street – energy performance model	130
7.15	Parmiter Street – coordination model	130
7.16	Parmiter Street – precast wall panels	130
7.17	Parmiter Street – wall panel installation	131
7.18	Pemberton remediation site	132
7.19	Precast concrete terraces	133
7.20	The new Parc y Scarlets	134
8.1	Rogers' innovation diffusion model	139
8.2	US labour productivity trends	173
8.3	UK construction productivity	174
9.1	World population projections	200
9.2	Vostock ice core data (temperature, CO_2, dust content)	202
9.3	Global warming observations 1880–2010	202
9.4	Earthrise	204

Foreword

Over the past two or three years, Building Information Modelling has become one of the most widely debated and written about topics in construction. Almost all of this discussion has focused on explaining what BIM is, what the benefits of using BIM might be, and how to use BIM techniques most effectively. In the short term, these probably are the issues the industry most urgently needs to grapple with. However, in the longer term – starting in about five years' time – a much bigger set of questions will come to the fore. These are to do with the way in which the industry responds to the capability of building with perfect information. The discussion becomes less about what we can do with BIM, and more what BIM will do to us.

When I started writing this book, I thought the most important thing to do was to point up the problems associated with drawing-based building design and to demonstrate how BIM could be used to overcome those problems. From the beginning, I had a strong sense that this use of BIM would have many powerful, beneficial effects on the industry, and that seemed to be the key issue to explore. However, as I started to appreciate the effects that advanced information technologies have had on other sectors of the economy, I saw more clearly what I now believe is the most important feature of Building Information Modelling.

These other industries – by one account, comprising nearly 80 per cent of the modern economy – have almost all been transformed by a process that might be called 'digitisation'. In most industries this has been a two-stage process. First, firms improved progressively and fundamentally the quality of the information used in the operation of their production processes and in the management of their businesses. Second, firms, sometimes acting as part of larger industry groupings, introduced fundamental improvements in their internal and company-to-company data exchange and communications processes.

In almost all cases, the changes to which I refer were the culmination of numerous relatively small, incremental steps, undertaken tactically, in response to specific local pressures or opportunities. There is no evidence that the firms in question set out with any sort of big, strategic goal. None had any prior idea of the nature or extent of the transformation they were initiating. Although each step taken at the time represented no more than a relatively minor change to current practice, the cumulative effects have been truly transformative: profoundly disruptive, massive in scale and often deeply traumatic in terms of their organisational and social consequences. Some observers have described this process as a 'digital revolution', similar in impact to the agricultural and industrial revolutions that shaped earlier eras of human history.

BIM is now beginning the same sort of process in construction. BIM systems generate fundamentally far higher quality information than drawing-based design systems are capable of doing. This improved information quality is already starting to change things in quite subtle ways. As the BIM authoring tools and data-exchange standards and communications protocols continue to mature, as BIM capability reaches a critical mass over the next five to ten years, the concept of end-to-end transmission of computable data throughout the industry's supply networks will gradually become a reality. This is digitised construction; building with perfect information. This form of construction will be as different to today's analogue industry, as today's digital manufacturing and retail industries are different to their 1970s analogue predecessors.

Inevitably perhaps, most people in the industry are leaving the BIM discussion to the 'techies': IT and CAD people. The aim of this book is very specifically to broaden the discussion of BIM futures. It is particularly to encourage non-technical people – business managers, teachers and policy makers – to participate. BIM is no longer a tactical, technical issue – it is far too important to be left to the technicians alone, as I'm sure most technology people would agree.

Acknowledgements

I would like to express my thanks to many friends and colleagues who have provided encouragement and support, both in preparing this book and also more generally. Mark Bew, Simon Rawlinson, Jeff Stephens, Nick Nisbet and other members of BuildingSMART UK helped greatly.

Martin Hewes of Hewes and Associates and Peter Fordham of Davis Langdon also provided key assistance. Between them, they compile some of the most interesting longitudinal data, and produce some of the most illuminating analyses of the UK construction industry currently available.

I would also like to thank many friends and ex-colleagues at the company formerly known as Bovis. These include particularly Mike Walker, Alan Crane, Hugh Coulter, Ian White, Peter Jacobs and John Spanswick, all of whom, at various times, provided essential support and encouragement.

The two pioneers of building information modelling who have personally most influenced and stimulated my interest in this subject are Jonathan Ingram and Jim Glymph. To both, my thanks and admiration.

Ray Crotty

Picture credits and sources

Figure 7.1 Source: R. Green, 2011, 'Just How 3D Are We?' *Cadalyst* magazine, February 2011. http://www.cadalyst.com/management/just-how-3d-are-we-part-1-13713 Retrieved 24 February 2011.
Figure 7.2 Courtesy Wikipedia. http://en.wikipedia.org/wiki/File:Weisman_Art_Museum.jpg Image by en:User:Mulad. Retrieved 12 December 2010.
Figure 7.3 Courtesy Wikipedia. http://commons.wikimedia.org/wiki/File:Barcelona_Gehry_fish.jpg Retrieved 10 December 2010.
Figure 7.4 Courtesy Wikipedia. http://en.wikipedia.org/wiki/File:Guggenheim-bilbao-jan05.jpg Photograph taken by User:MykReeve on 14 January 2005. Retrieved 10 January 2011.
Figure 7.5 Courtesy Wikipedia. http://en.wikipedia.org/wiki/File:Guggenheim Bilbao.jpg Retrieved 10 January 2011.
Figure 7.10 Courtesy Wakefield Council. http://www.waterfrontwakefield.com Retrieved 18 January 2011.
Figure 7.11 Courtesy Wakefield Council. http://www.waterfrontwakefield.com Retrieved 18 January 2011.
Figure 7.18 ©URS Corp / David Lawrence http://www.urscorp.eu/projects/project.php?project_id=622 Retrieved 14 January 2011.
Figure 8.2 C. Eastman, P. Teicholz, R. Sacks, and K. Liston, 2011, *BIM Handbook; A Guide to Building Information Modelling,* 2nd edition, Chichester: Wiley. Figures 1–4 Indexes of Labour Productivity for Construction and Non-Farm Industries, 1964–2008. Adapted from research by Paul Teicholz at CIFE.
Figure 8.3 Source: Constructing Excellence, *2010 UK Industry Performance Report*, London: Constructing Excellence, p. 11.
Figure 9.1 Source: Population Division of the Department of Economic and Social Affairs of the United Nations Secretariat, *World Population Prospects: The 2007 Revision, Population Database.* http://esa.un.org/unup, Retrieved 10 November 2010.
Figure 9.2 Source: Wikipedia http://en.wikipedia.org/wiki/File:Vostok_Petit_data.svg. Retrieved 12 December 2010.
Figure 9.3 Source GISS/NASA. http://data.giss.nasa.gov/gistemp/graphs/ Retrieved 2 December 2010.

Figure 9.4 Source: NASA. The Earth from Space, taken by Apollo 8 crew member Bill Anders on December 24, 1968. http://upload.wikimedia.org/wikipedia/commons/thumb/a/a8/NASA-Apollo8-Dec24-Earthrise.jpg/600px-NASA-Apollo8-Dec24-Earthrise.jpg Retrieved 10th November 2010.

1 Introduction

1.0 Introduction: problems with drawings

When one is immersed in it, working in it every day, it can be difficult to stand back and appreciate just how information-intensive the construction industry is. Modern buildings[1] are amongst the most complex things we create, and the teams required to construct them are amongst the most complicated forms of human organisation. Even on relatively small projects, this combination gives rise to a virtual storm of information: thousands of individual documents – many in a state of continuous revision – circulating rapidly amongst a large, transient array of very different types of individual people and firms.

There are two key challenges in trying to cope with this situation: the quality of the information being generated and used on the project, and the means by which this information is communicated and shared amongst the project team.

There have been countless official and semi-official reviews, investigations and reports on the performance of the construction industry over the past 100 years or so. Almost all of these have pointed to poor standards of information management as being, in one way or another, instrumental in the industry's under-performance. And, although most have suggested fixes, largely of an organisational or contractual nature, none has been able to offer a real solution. Until now, there has been very little that the industry could do about these issues; but with Building Information Modelling (BIM), things may be about to change.

Most of the information used on a construction project originates in the architectural drawings created in the course of the design process. Drawings – even when they are created using CAD systems – are notoriously poor containers, or conveyors, of information. There are two main problems.

First, drawing-based information is inherently untrustworthy; anyone who receives this sort of information cannot assume that it is true. Instead, before using

---

1 The effects discussed here are not limited to buildings; they include almost all forms of built facilities, including roads, railways, process and petrochemical plants and so on. The word building should therefore be treated as an abbreviation for all these other constituents of the built environment. There are also elements of this overall argument that apply strongly to the operation and maintenance phases of the building life-cycle. Though not addressed explicitly throughout the book, these aspects will be highlighted where relevant.

it, he or she must check to ensure at least that it is clear, consistent, coordinated and correct.

To carry out these checks effectively and consistently takes time and requires high levels of skill, discipline and judgement – qualities not always plentiful on fast-moving projects.

The second major problem with drawing-based information is that it is essentially un-computable; anyone who receives such information and wishes to reuse it for computing, must first decode it, then – usually manually – re-enter it into his or her own system. This is a hugely wasteful activity which introduces a whole new set of errors into the project information flows.

These points are not intended as criticisms of designers or of the techniques they use. The problems are simply unavoidable in drawing-based design; they are inherent in the way in which drawings are created and managed. However, their ultimate effect is to lock construction into a craft-based mode of operation, a way of doing business that would be quite recognisable to medieval builders and their clients.

BIM promises to break that lock, both by improving dramatically the inherent quality of building design information and by improving dramatically the mechanisms and procedures by which information is communicated and shared amongst the members of a project team. It helps to think of the BIM approach as comprising two discrete facets:

- one or more modelling systems in which the actual building design is carried out; and
- a supporting set of data-interchange standards and protocols by means of which the individual models communicate with each other, and with other applications.

The major product and mechanical manufacturing industries moved on from craft production, initially to mass production in the early twentieth century. Towards the end of the century, mass production was superseded by lean production and mass customisation modes of operation. Sophisticated quality assurance techniques, just-in-time methods, and other innovations have contributed to these developments. But arguably the two most important achievements of these industries, in this context, have been the replacement of drawings with models, as the basis of the design process, and dramatic improvements in the integration of information flows throughout their value chains.

The result has been a dramatic transformation in almost all areas of mechanical and product manufacturing, leading to a form of operation called computer integrated manufacturing (CIM). In this mode of working, conventional products can be produced to far higher quality, far less expensively than in the past; and many new products, which would have been impossible to make using earlier techniques, are now commonplace.

BIM does very much the same for buildings. It can be used to produce conventional rectilinear buildings, of greater variety, higher quality, more efficiently and more economically than conventional methods. But, as demonstrated by the Guggenheim Museum in Bilbao and many other buildings by Gehry Associates, the Gherkin in London by Foster & Partners, the National Stadium (The Bird's Nest) in Beijing by Herzog & de Meuron, and other projects, modelling systems can also be used to produce buildings of dazzling complexity, structures that simply could not have been contemplated using drawing-based methods of architectural design.

It would be misleading to imply that the same sort of dramatic change might be about to happen in construction as has happened in manufacturing over the past two decades, but it is no exaggeration to suggest that the industry is on the threshold of a major, fundamental change in its basic methods of operation. This won't impact all forms of construction equally or simultaneously, but there is almost no area of the industry that in, say, ten years time, will remain unaffected by the changes discussed here.

1.1 BIM modelling systems

The BIM modelling systems currently in use all involve the designer assembling a computerised model of the proposed building in virtual 3D space, using intelligent components, inserted at precise orientations, into precise locations in this space. The individual components in BIM systems are organised into classes or families of objects that correspond directly with classes of building components in the physical world; modelled walls, columns, doors, windows and so on, all have direct real-world counterparts. And the process of building the model – inserting the components – is in itself analogous to the processes involved in building the physical building; there are specific rules governing the order in which components may be inserted, where they may be placed and how they interact with each other, just as there are rules governing construction operations in the real world.

The BIM system vendors provide libraries of precisely specified component families covering all of the standard components encountered in building projects. They also provide methods whereby users can develop their own components to suit their particular needs, while adhering to the particular vendor's data specification. Components are assigned properties which capture in the computer the characteristics of the objects they represent. These can include physical features such as geometry, density, modulus of elasticity, thermal capacity and so on, as well as economic attributes such as vendor details, unit cost, delivery lead time etc.

The properties assigned to components can also include rules which control how they behave. For example, the width to height ratio of a particular type of window might be set to equal 2/3. If the user changes one of the dimensions of such a window, the second dimension will adjust according to the 2/3 rule. Component properties can also be used to create connections or other forms of relationship between components. Thus a rule could be set relating two components, such that if the diameter of one component is changed, the radius of a hole in the related component will also change, as determined by the rule.

4 *Introduction*

The rules and other properties of the components are defined in the modelling systems using parametric equations. The systems vendors determine the parameters to be used for each family of components, and the permissible range of values for each parameter. The user selects the particular parameter value he or she requires by selecting from a menu, using a slider control or some such method. Using these capabilities, a great deal of knowledge of design and construction best practice can be embedded by the system vendor in the selection of parameters used and by the designer in his or her choice of parameter values.

A final point to note is that whenever a change is made to a particular component in these systems, its effects are transmitted to all of the components to which the initial component is connected or related. The details of the change, the name of the user making the change and the time and date may be recorded against every affected component. This enables the design, at any point in time, to be rolled back to any earlier point in its development, which provides for very rigorous design change control. More importantly, it also allows the design team to carry out large numbers of what-if explorations of the full range of design solutions available.

So, the most important capability of an individual BIM modelling system is that it enables the designer to 'build' the building in a computer-generated virtual world, before going to concrete, so to speak. He or she assembles the model using 'intelligent' virtual components, each of which is exactly analogous to a building component in the physical world. The building can be viewed from different angles and in many different ways. Many aspects of the building's behaviour can be tested in detail, and design changes can be implemented quickly and confidently. Accurate drawings of the building and definitive schedules of the components that make it up can be generated easily. Data files can be provided for use in computer numerical controlled (CNC) manufacturing of components. And the building's construction and operation can be simulated in very precise detail, all before procurement and construction of the real building commences.

This is what might be possible if a single designer, using a single system, could design a complete building. However, generally speaking, this type of idealised BIM is not possible. Modern building projects involve large numbers of designers and construction firms, many of whom provide specialist inputs at different points in the overall design and construction process. Nowadays, most of the firms who do this carry out their technical and design work using computer systems of one sort or another. Some of these are intelligent, component-based BIM modelling systems in the sense outlined above; others are conventional 3D CAD systems, some are just 2D drawing systems, and some are non-graphical, analysis, simulation or business systems.

1.2 BIM and standards

The second part of the BIM approach therefore comprises the communications, or data exchange, standards and protocols necessary to enable all these different types of systems to talk to each other. The data standard is based on the idea of a *lingua franca*, a common language, that the relevant systems can all speak, or more accurately perhaps, a language that they can all read and write. As indicated above, the central construct in the BIM approach is an intelligent, component-based, 3D model of the building. All of the currently available BIM modelling packages operate at the component level of detail, and they all provide comprehensive libraries of components with which the user can build his or her model.

So the starting point for the data standard is the 'component', both as it exists in the real world and as it is represented in the different modelling systems. A technical body, called the International Alliance for Interoperability (IAI) was set up in 1995, with representation from the worldwide construction industry and the major CAD vendors, to pursue this concept. Since then, the IAI has developed a comprehensive set of component specifications, called industry foundation classes (IFCs), which provide a neutral, systematic description of all of the main families of construction industry components. Any IFC-compliant computer system is required to be able to output a large proportion of the information it contains about any given component in this neutral format, and equally, any similarly compliant system should be able to import those details and use them in its internal working – without the need for intervention on the users' part.

There are unavoidable limitations in the completeness of the sets of information that can be exchanged between modelling systems. Thus, basic information, like the geometry, location, materials properties, cost, delivery date and so on can usually be exchanged fairly readily. However, many of the attributes of components that make them 'intelligent' are specific to the originating system, and are difficult to translate. This includes properties like the way objects connect to each other, and the way in which components embody knowledge of construction practice. However, these 'intelligent' properties are generally discipline-specific, in the sense that, although they may be interesting to view or read, they do not usually need to be edited, or otherwise reused by firms other than the originators of the components in question.

The IAI – recently renamed buildingSMART – has also been working on the development of a set of procedures and protocols that can be used to determine what information, in what format, should be communicated between particular firms, at particular points in the development of a project. These protocols are not yet complete, but, as they can usually be agreed on an *ad hoc*, project-by-project basis, they are arguably not as critical as the data standards. The key thing to recognise is that although the standards-setting processes are not yet fully complete, the standards that are available are workable and cover most of the issues that matter. As always in this area, an open, pragmatic approach yields most of the benefits that an ideal, perfect solution would provide, for a great deal less cost and effort.

An important point to note here is the fact that, almost as a by-product of the modelling standards work, the IAI has actually created the basis for a building component classification system that could be used for data interchange between many relatively mundane construction-related applications: estimating, planning, accounting, production management and so on. This opens up the possibility of managing construction at the component level, injecting precision and effectively eliminating the need for individual human judgement in this difficult area. The importance of this development will emerge over time.

Thus in deploying the two aspects of BIM we achieve two hugely important improvements:

- far higher quality design information than with conventional tools, through the use of intelligent, parametric, 3D building models;
- far higher quality, more efficient communication amongst the systems of the project team members through the use of clear and effective data exchange standards.

Over the past 20 or 30 years the aeroplane, car, consumer product and other manufacturing industries have all started to embrace their own equivalents of these two developments. The combination of model-based design, and data-integrated supply chains has been revolutionary wherever it has been applied. In all reported cases, dramatic improvements have been achieved in product quality and variety, in time to market, in reduction of work in progress and in many other aspects of their operations.

However, most of the manufacturing industries in which BIM-type methods have been employed to date have been dominated by relatively small numbers of large manufacturers. Rather than focusing on public standards, these firms have tended to use their market power to coerce their suppliers into conforming with their proprietary design and data-exchange standards. They have overcome the data communications problem by imposing private, application-specific standards.

Construction, however, is a notoriously fragmented industry, with very low levels of market concentration. So the Ford or Boeing model of mandated proprietary standards will not work. Instead, more public, non-proprietary standards will have to be deployed. Fortunately, IAI has published a number of iterations of its data exchange standard, the industry foundation classes (IFCs), and all of the key vendors can now import and export models in IFC format.

In the retail industries, a combination of technologies and data standards – electronic point of sale (EPOS) systems, together particularly with the universal product code (UPC) data standard – enables communities of retailers to achieve fully digital, end-to-end flows of data along the most diverse and extensive supply chains imaginable. Retail has been transformed as a result.

It may take some time, but with the thoughtful support of the systems vendors, continuing work on the IFCs and other standards, and an energetic level of informed discussion in the industry, BIM can be expected to have a similarly transformative effect on construction.

1.3 BIM in action: Ikea kitchens

That, very briefly, describes what BIM is. To get a greatly simplified idea of the capabilities of these systems, the reader might go to www.ikea.com and download one of their online room planning tools – the kitchen planner for example. Use this to create the walls of a kitchen space in plan view. Now add the doors and windows and note how they seem to 'know' what they are, how to connect to each other and how to display themselves in different views. Then add Ikea's intelligent product components: floor and wall cabinets, sink units, electrical appliances and so on. Move them around as you wish, again noting how they obey various rules about how to behave. To complete the design, apply finish details, different door faces, different worktops to suit. Check how they look in 3D; rotate the model so as to see it from different perspectives. Print off pictures, plans and elevations as required. That's the modelling part of BIM.

When the initial design activity is completed, Ikea's information management systems kick in, to pass the information generated by you, the customer, to the relevant business functions and to the company's partners, as appropriate. So, to finish, print off a list of the components installed in the virtual kitchen. This will automatically reflect the latest in-store prices. Continue to adjust the design until the layout, contents and cost are satisfactory. Then click the button to place the order on-line. The system checks the availability of the various components, passes a request to production for out-of-stock items and debits your credit card. Do you want the kitchen installed for you? If so Ikea will introduce you to an installation company and will give you a price, based on a standard installation rate for each component, all agreed in advance with the installers. Do you want the items to be delivered? If so, Ikea will pass the necessary information to their logistics partner and will advise you of their agreed consignment date. You just sit back and await the arrival of the lorry …

This may seem a flippant way of introducing one of the most important developments in construction in recent decades, but note what's going on here. The Ikea design system encapsulates very effectively the general power of intelligent, parametric, component-based, 3D modelling systems. The associated business systems, such as stock control, production management and billing all make direct use of the data created in the design process. One might think of the Ikea system as being essentially a user-friendly kitchen modelling tool acting as the front end to a complete kitchen manufacturing and supply process. Scale up the information management capabilities of this little tool to one of the latest full strength building modelling systems and it becomes clear just how transformative this technology could be.

1.4 The advantages of model-based design

The ability to build the building in a model and to completely test and analyse the design, prior to construction, adds up to a dramatic improvement in basic design processes. However, the really transformative power of these sorts of intelligent models lies in the remarkable improvement in the quality of design information that they produce. Conventional, drawing-based design documentation suffers from four main deficiencies:

8 *Introduction*

- the use of arbitrary lines and symbols leads to ambiguity and misunderstanding;
- it can be difficult to ensure that individual document sets are properly internally consistent;
- it can also be difficult to ensure that related document sets are correctly coordinated;
- it can be difficult to ensure that the documentation is fully complete.

To repeat, none of this is intended as criticism of designers working in construction; they do as well as can be expected with the tools available. The problems listed are simply inherent in the drawing process and in the supporting techniques used in conventional building design. Until now, there has been no other way of getting the design ideas out of the architect's head and into the hands of the craftsman doing the building work. The industry has lived and struggled with this problem for centuries. It has depended on the application of almost heroic levels of individual experience and judgement on the part of clients, engineers, builders and tradesmen to overcome the inherent deficiencies in the documents and to deduce or guess their true and full intent.

These problems should not occur in a well-managed, model-based design project. The sheer quality of the information generated in BIM models completely changes the nature of the process. The information provided to the client, to fellow designers and to contractors is of a fundamentally different nature to that generated by drawing-based processes. Model-based information is essentially as good as the information generated by product designers in manufacturing industries. It's inherently unambiguous, fully internally consistent, accurately coordinated, and complete. And above all, it's entirely computable. That is to say, the data generated by one of these systems can be used, without any need for human intervention or interpretation, as direct input into other computer systems. It's so accurate and complete that it can be used directly to drive computer numerically controlled (CNC) machine tools for carrying out such operations as milling, cutting, punching, boring and shaping materials like steel, aluminium, timber and even concrete.

In summary, the model created using the latest BIM tools is a powerful, clear, flexible and rich representation of the designers' intentions. It offers enormous benefits to all the key players: the client, the architect, the members of the larger design team and the contractors. The most important of these are the following.

1.4.1 *The client*

In the BIM approach, the design can be presented to lay viewers, such as client organisations, as a photo-realistic, walk-through – 'what you see is what you get' – model. This can include aspects of the building's sustainability, energy performance and so on. This explicit, non-cryptic method of representation greatly improves the client's confidence in his understanding of the scheme and enables early decisions to be made with much greater certainty than is usually the case.

Subsequent stages of the project can then proceed smoothly, with a minimum of client-instigated design changes.

The most important advantage of BIM to the client will be a dramatic improvement in certainty of outcomes. In a BIM-based industry, every significant aspect of the building's construction and performance will be modelled and tested fully, before manufacture and assembly commence. BIM models create and output information of the same sort of quality as that used in the manufacture of modern consumer products, for example. Just as those products are guaranteed by their manufacturers, it will become possible for the designer/manufacturer/assemblers of BIM-designed buildings to provide buildings whose cost, delivery and in-use performance can be fully guaranteed, for whatever period is deemed appropriate.

1.4.2 The individual design firm

Individual design firms benefit from BIM working in two main ways. First, because BIM models are inherently internally consistent, the need for tedious checking across different documents is greatly reduced. And, because drawings can be generated automatically from the model, the proportion of the overall effort going into the production design phase of the project is significantly reduced. Proportionately more effort can be focused on the creative, problem-solving aspects of design. So the proportion of higher grade work, thus higher value work in the firm's package of services increases. Because this work happens relatively early in the design process, it also has the effect of moving the resource peak – and therefore the cash flow peak – forward in the programme.

1.4.3 The project design team

Compared with drawing-based design, it is also relatively easy to coordinate the design contributions of different disciplines by incorporating them into a shared BIM reference model. Amongst other things, this means that many fewer multi-disciplinary design review and integration cycles are required to complete the design of the building, leading again to a much more efficient overall design process. It also means that the incidence of construction clashes on site can be more or less eliminated, resulting in large construction cost savings.

1.4.4 The contractors

For the contractors, the obvious benefit of a BIM design is the ability to visualise the building in great detail and to simulate its construction. However, of potentially greater value is the ability to generate from the model complete and definitive statements of the scope of work of the overall project or of individual work packages. This has two particularly important implications: first it makes it possible to tender trade contracts more fairly and more competitively. It does this by removing the need for bidding contractors to allow for scope risk in their prices. It also removes both the need and the opportunity for contractors to bid uneconomically low in pursuit of claims.

The second area in which a BIM model benefits the contractors is in planning, cost-control and other project-management activities. The key here is that the definitive scopes of work generated by a BIM model enables much more robust cost and schedule targets to be established at the outset of the project, and much more accurate and systematic progress assessments to be made during the course of the work. The result is that forecast end dates and out-turn costs will be much more accurate and reliable. A lesser, but still significant advantage of BIM-based project management is that it enables greatly improved labour performance and productivity data to be captured for reuse on subsequent projects – it enables companies to learn.

1.4.5 Building owner/occupier

The principal advantage of BIM to the owner/occupier is that an 'as-built' BIM model can be used very effectively to support intelligent operation and maintenance of the building. Actual versus designed energy usage can be monitored. Reconfiguration and refurbishment exercises can be planned in great detail and communicated easily to building users. Health and safety aspects of the building's operation can be supported accurately. The 'as built' model allows the owner to simulate, test and generally optimise the functionality and performance of the building throughout its life-time. It becomes a powerful asset management tool which enables the owner to truly maximise the return on his investment in the building.

1.4.6 Summary

A BIM model can overcome most of the most serious failings of conventional drawing-based design: greater client certainty earlier; improved consistency and easier coordination of design documentation; improved, complete procurement documentation; much more powerful construction and project management tools; and much more valuable 'as built' and record information for the owner. The result will be substantially more profitable firms of all types in the sector delivering projects a great deal more reliably.

1.5 The bigger picture

The fact is that what we call BIM today is just the leading edge of a wave of innovations that, in the coming years, will transform construction, in the same sort of way that the manufacturing and retail industries have been transformed by computer integrated manufacturing (CIM) and electronic point of sale (EPOS) systems over the past 20 years or so. EPOS in retail actually provides a powerful analogy for our purposes.

The essence of EPOS is that it enables retail operations to be driven by data – information in a trustworthy form that never needs to be checked. From the point at which its barcode is applied to a product item on the farm or in the factory, to the point at which the customer passes it through the checkout, the data generated in the life-cycle of any given item is managed across the entire supply chain, entirely automatically. The information about the item flows seamlessly from one end to the other along the supply chain. There is no need to check, or for any other form of intervention or human judgement. And by eliminating the need for judgement and human intervention from the supply chain, retailers eliminate most of the errors that manual business processes are prone to.

The really important long-term benefit that BIM offers is the potential to manage all aspects of the life-cycle of a building using the same sort of trustworthy information – computable data. At present, from the point at which an element first appears on a designer's drawing to the point at which its real-world counterpart is installed into the real building, all the information, every transaction and every event in its life-cycle is recorded and managed manually. Every transaction requires human intervention. Every piece of paper, every CAD drawing and printout must be checked, at least once, before being acted upon or reused.

This is of the nature of untrustworthy information – it cannot be taken as being true without being checked and validated. Besides being enormously wasteful, every one of these checking and validation exercises is in itself a compounding source of error. BIM on the other hand, generates inherently trustworthy, computable information. It offers the possibility of more or less completely automating these transactions, eliminating the waste they introduce and the errors they generate. The end-to-end stream of BIM data will help unify the industry's supply chains, freeing construction from its craft origins, transforming it into a modern, sophisticated branch of the manufacturing industry.

BIM may have significant impacts on other aspects of the construction industry's operations, such as off-site manufacture and prefabrication of building components. BIM will accelerate the adoption of off-site working in two ways. First, because the on-site construction processes will be far smoother and more predictable than at present, it will be easier to integrate or connect the two areas of activity. Much more efficient use of production line techniques will be possible. BIM can provide design information in the form of computerised data, which can be input directly into computer-controlled machines. This again will increase the efficiency of factory-based production lines.

The more productive off-site manufacturing becomes, the more it will be used and the greater the value of individual manufactured items. The higher the inherent value of these individual items is, the lower proportionate transport costs will be. This in itself will increase the size of the market for such items. International (initially EU-centred) standardisation will add to this effect, so that in a relatively short timescale, the market for a wide array of construction components will become global.

The concomitant of increased off-site manufacture is reduced demand for engineering, supervisory and craft skills at site level. Like mainstream manufacturing, the need for craft-based modes of working will disappear from construction.

One can anticipate these developments simply by observing the experience of other industries and economic sectors, as they became digitised. There will be many other advances, possibly more significant ones, that cannot yet be envisaged. But the direction of evolution is definitely towards a safe, high-quality, predictable and, for the survivors, a highly profitable construction industry.

The main purpose of this book is to try to capture the depth and extent of the impact that BIM will have on the UK construction industry. Thus Chapter 2 reviews the strategic background of today's industry. Out of the dozen or so key performance indicators (KPIs) identified by the UK Constructing Excellence project, the book focuses on two: the chronic failure of the industry to deliver projects predictability; and the industry's impossibly low level of profitability. These are the two fundamental problems of construction. And as will be shown, they are both caused mainly and directly by the poor quality of information provided by conventional design techniques.

Chapter 3 reviews in some detail the features and impacts of drawing-based design information. It considers the difficulties of dealing with this in the design, procurement and construction phases of projects. Chapter 4 considers the way in which the BIM approach might be expected to overcome these difficulties, again in all three of the main project phases.

Chapter 5 reviews the history of computer-aided design (CAD) technology. The key contribution of British designers and software developers to the creation and development of CAD and particularly component-based 3D modelling is highlighted. Following on from the story of their development in the CAD industry, Chapter 6 discusses BIM tools in greater detail and introduces the information management and communications standards and protocols that are essential to making BIM work properly in the multi-player project environment.

BIM is a very fast-moving technical approach to building design and construction. Chapter 7 attempts to capture the current status of BIM deployment in the UK, and also in western Europe and the USA. This takes the form of a brief review of some recent user uptake surveys, accompanied by a selection of short case studies of project teams who have actually used BIM or near-BIM techniques in earnest.

Construction is a particularly information intensive industry. Conventional applications and communications technologies have hardly scratched the surface of this challenge. BIM enables the industry to address the problem of information intensity directly, much as other technologies have enabled other industries to tackle their particular information challenges. In Chapter 8, the experience of industries like banking, manufacturing and retail are sketched out and the disappointing history of construction investment in IT to date is outlined. These sections provide the background against which the probable trajectory of the construction industry is projected in Chapter 9.

In a sense the whole purpose of this book is captured in the final chapter. Building Information Modelling generates effectively perfect information. Chapter 9 discusses this phenomenon and describes its transformational impact on the future construction industry – which will be a truly global industry.

The main aim of the book is to promote well thought-out, pragmatic implementation of BIM-based approaches to construction. The book is aimed primarily at the people who make the decisions that directly influence the organisational shape and overall execution plans of significant projects: project directors, project managers, design team leaders and construction managers. Although it does address the practicalities of implementing BIM, it does not aim to be a BIM 'how to ...' book. And although it discusses the technical aspects of some BIM tools, it is not a book about BIM technologies *per se*. It's a book about construction and about how BIM will change construction; a subtle distinction, but an important one.

Anyone with a background or serious interest in the construction industry should find something of interest, hopefully something thought-provoking, in this book. It has its basis in the UK industry – a true laboratory for industrial innovation, whatever its detractors may say. Its UK bias however, does not mean that the book disregards other countries and other readerships. Each country's industry must learn its particular lessons in its own local way; hopefully this account of the UK experience will help inform discussion of their own industries by readers in other countries.

2 UK industry background

2.0 Introduction

The purpose of this book is to explore the likely impact of Building Information Modelling on the UK construction industry. Later chapters will address the operational aspects of BIM implementation: reasons for doing it, the conditions necessary to do it successfully and the anticipated results of doing it, both at the level of the individual project and at overall industry level.

The first part of this chapter provides a brief review of a number of background features of the industry that are likely to have a significant bearing on the adoption of the BIM approach in the UK. The second part discusses the poor record of predictability and profitability in the industry and outlines in broad terms how BIM might help resolve these problems.

2.1 Key background features of UK construction

There are four issues or features of the industry that are of interest here:

- industry structure: ways of looking at construction, identifying decision makers;
- the industry's history of self-analysis and its results;
- changing roles and relations in construction; and
- the industry's capacity for innovation.

These four issues are closely interwoven in most discussion of the industry, but it's necessary to disentangle them to some extent, in order to see clearly how they will influence the dissemination of BIM over the coming five to ten years.

2.1.1 Economic structure – decision makers

Although the UK construction industry comprises an enormous number of firms, only a tiny proportion of these are of any substantial size. For example, although in 2008 there were 53,500 registered contractors in the UK industry, only 283 of them employed 300 or more people. These larger companies, although they constitute only 0.14 per cent of the firms in the industry, employed 24 per cent of the total construction workforce and generated 35 per cent of the industry's output.[1] A slightly different perspective on this issue is given in *Building Magazine*'s annual league table. The latest version shows that 30 of the top firms accounted for 80 per cent of the turnover of the top 75 UK contractors.[2] So the economic mass of the industry is heavily concentrated in a relatively small number of relatively large firms.

In trying to deal with something as diverse and as complex as construction it is natural that people should create mental models, simplified versions of reality, that they can use for analysing and solving problems or for predicting future developments. In construction, most analysts and policy makers build their models around this top layer of the industry. With a membership of 200 to 300 firms, it comprises a manageably small number of individual entities that are relatively homogenous in their structures and can be expected to behave relatively consistently. Assuming that whatever is true at this level will also be true further down the system, they use these models to draw conclusions about or to apply leverage to players throughout the industry's economic hierarchy.

In fact there are many ways in which these types of industry models can fail: what's true for players at the top is not necessarily true further down, and so on. But a more fundamental flaw with this approach is that it places the construction companies at the centre of the construction industry. The true centre of the industry, the place where value is added and innovation is forged, is on projects, not in corporate head offices. True, projects are transient, fast-moving things, elusive and difficult to comprehend from the outside. Nonetheless, to properly understand how innovation comes about in construction, one's picture of the industry must have the project at its centre.

This model focuses on the fact that in 2008 a total of 53,500 new construction contracts were awarded in the UK. Of these, 6,901, 13 per cent of the total number, comprising 77 per cent of the total value, were projects worth £1m or more. And 218 projects, 28 per cent by value of all the new work awarded, were worth £20m or more. This pattern has remained consistent over the past five years.[3]

This looks very much like the same sort of hierarchical analysis as the companies-based one outlined above: a top layer comprising a manageable 200 or so entities, a middle layer of a few thousand and a very large base layer of tens of thousands. But this is much more useful for our purposes. To repeat, most innovation in construction takes place on projects. The overall complexity of buildings tends to increase with size, and project organisations become more complicated very rapidly as the number of participating firms increases. Innovative use of tools and techniques aimed at managing this complexity is therefore more likely to happen

---

1 Office of National Statistics(ONS), *Construction Statistics Annual 2009*. Tables 3.1, 3.3, 3.4.
2 Hewes Associates, 'Construction Industry League Tables', *Building Magazine*, 31 July 2009.
3 ONS, Table 1.6.

on larger projects than on smaller ones. This has been the case so far with IT systems; BIM will almost certainly follow a similar path.

So, given that about 200 large projects are started every year, and assuming an average project lasts about three years, there are about 600 large projects at various stages of development in the UK at any point in time. The particular people involved in a given project will vary over its duration, but assume there are on average five or six key people involved at any point. (At different stages of the project these might include: client's project manager, cost consultant/PQS, job architect, structural and M&E design team leaders, project managers from main contractor, structural frame contractor, cladding contractor, M&E contractor.)

This rough calculation suggests that there are about 3000 to 4000 individuals in senior positions directly involved in the design and delivery of major projects in the UK at any point in time. These are the most important people in the industry in many respects. They are certainly the people who will most strongly influence the dissemination of BIM methods through the industry. They are the people at whom this book is mainly directed.

2.1.2 Self-analysis and its results in UK construction

Most people who do business either in or with the modern construction industry become aware pretty quickly that all is not well in construction. The industry has a high profile, but low esteem. As an important sector of the economy its performance is a matter of national importance. As a result, the industry is subjected to regular official and semi-official investigations. These have been – or perhaps just seem to have been – particularly frequent over the past 60 years or so. Murray and Langford[4] provides a useful summary of the most important of these exercises. It's an important document, and it makes pretty depressing reading. In summary:

- The reports all identify the dislocation between design and construction as a key problem; they urge closer integration within design teams and between designers and constructors.
- They identify short-termist thinking as a strategic industry problem, and urge major clients, government and quasi-governmental bodies in particular, to provide long-term continuity of work to the industry.
- They identify uncoordinated, incomplete design information as a common cause of poor construction performance. The tendency of client teams to want to rush to site is noted.
- They identify lack of management skills as a further cause of poor performance on site.

---

4 Murray, M. and Langford, D., *Construction Reports, 1944–98*. Oxford: Blackwell Science, 2003.

- They propose innovative modes of project organisation and novel forms of contract as a way to overcome these problems.

Perhaps the most depressing thing to note is the extent to which the reports seem to repeat each other, over and over again throughout the period, both in their analysis of the industry's problems and in their recommendations for improvement. But nothing really seems to change.

In fact, it's difficult to imagine how something like an industry – an entire sector of the economy – might go about solving these sorts of problems. But at least construction tries. As Murray and Langford demonstrates, the industry has shown repeatedly that it is aware of its failings and that, in some sense at least, it is anxious to improve.

Thus, in the wake of the latest big review – the Egan Report[5] – government and the industry jointly established a body called Constructing Excellence (CE), to identify, measure and analyse the causes of poor performance in the construction industry.[6] At the time of writing Constructing Excellence has been at work for over a decade, monitoring a wide variety of industry performance indicators. Although a few of the key targets *Rethinking Construction* set for the industry have been achieved, most have proved elusive.

The most important single recommendation to emerge from Egan and subsequent initiatives was that the industry should attempt to reduce confrontational attitudes amongst its players and should instead embrace collaborative methods of working. Partly as a result, over the past 10 to 15 years the UK industry has been something of a laboratory for strategic and project partnering, for the use of frameworks and other non-confrontational approaches to procurement. UK industry experience in this regard has been exported to many countries around the globe, and a number of important public- and private-sector clients here and elsewhere have created deep and enduring collaborative arrangements with their project delivery teams.

This experience will be highly valuable as BIM usage takes off in this country. For example, although implementation of BIM is currently far more advanced in the USA than it is in the UK,[7] the use of collaborative forms of contract in the American industry lags behind British practice. But, leading individuals and organisations in the American industry have realised that BIM can be implemented in collaborative project organisations much more effectively than in those based on traditional, lump sum, competitive tendering. One result is that the American industry is now developing its own collaborative approach which they call integrated project delivery (IPD). Standard forms of contract and the

---

5 Construction Task Force, *Rethinking Construction*, The Egan Report. London: HMSO, 1998.
6 www.constructingexcellence.org.uk
7 According to McGraw-Hill: 'Almost 50 per cent of the US industry is now using BIM; all of those BIM users plan significant increases in their use; and the vast majority are experiencing real business benefits directly attributable to BIM.' *SmartMarket Report, Building Information Modelling (BIM) 2009*. Bedford, MA: McGraw-Hill Construction,.

necessary procedures documents have already been developed to support project teams attempting to carry out projects in this more collaborative way.[8]

There are two points to take from this. First, it has to be noted that few of the industry reports discussed in Murray and Langford resulted in sustained improvement in industry performance. However, the very fact that they took place and the amount of effort and goodwill invested in each of them demonstrates a genuine keenness to improve and a willingness to embrace fundamental change in order to do so. And, second, as the Americans are finding, the UK work on collaborative approaches will almost certainly pay off in the implementation of BIM methods in the coming years.

The people who have thought longest, worked hardest and invested most in these earlier attempts at industry self-improvement are the people discussed above – operations-level project leaders working on larger projects. Hopefully BIM will inspire them to renew those earlier efforts. Without their active involvement in its deployment, BIM will remain just another form of CAD; nothing fundamental will change.

2.1.3 Changing roles and relations

There are a number of pre-existing, long-term trends which will shape the development of BIM in the UK. Despite its reputation for conservatism and inertia, in recent decades, the industry has actually changed quite dramatically, both in the way it carries out projects and in the way it organises itself. Forty years ago the main contractor on most large projects self-performed the great majority of the work; today the main contractor simply procures and coordinates construction services supplied by a wide range of specialist sub-contractors.

This seems a novel development but is in reality just a reversion to the sort of arrangement that existed before the great building boom brought about by the Industrial Revolution. As Satoh describes it, the concept of the general contractor carrying out a complete building project, using his own labour force and equipment, on a fixed price, lump sum basis, became normal only in the mid-nineteenth century.[9] Prior to that, individual craftsmen or small teams of craftsmen were hired – trade by trade – directly by the client or his architect.

However, the Victorian era saw a huge surge in demand for construction, from local and central government, as well as from industry and commerce. General contracting evolved as the best, or as the least bad, way of meeting this demand. And as Morton points out, despite a few dramatic setbacks, it worked fairly well through the second half of the nineteenth century and through two world wars and the enormous demand for reconstruction that they generated.[10] However,

---

8   See for example: American Institute of Architects, *Integrated Project Delivery: A Guide*, Version 1, 2007.
9   Satoh, A., *Building in Britain – The Origins of a Modern Industry*. Aldershot: Scolar Press, 1995.
10 Morton, R., *Construction UK: Introduction to the Industry*. Blackwell: Oxford, 2002. Chapters 5, 6.

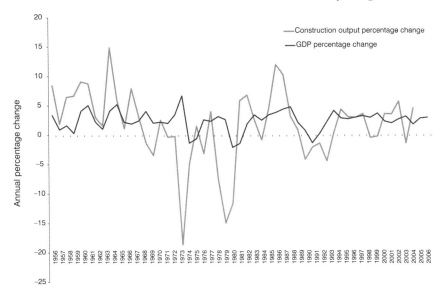

Figure 2.1 Construction output and GDP 1956–2006

general contracting is capital intensive and very risky. The 'old-fashioned' family-based contracting firms depended on the trust created by tight networks of close personal relationships within their organisations to manage this risk. In many cases these networks revolved around personal loyalties to the members of the founding families.

All was fine with these companies as long as they remained reasonably small. However, as they grew, they encountered two problems. First, through sheer size and through the progressive loss of interest on the part of younger generations of the founding families, the old trust networks became attenuated and less effective – greatly increasing problems of risk. And, as the families withdrew from running the businesses they also sought to disinvest in them. This, together with the need for finance to fund growth meant that the professional managers who remained, now had to resort to outside sources.

The timing of these developments, mainly in the 1950s and 1960s, coincided with an extended period of great volatility in the construction industry. The extent of this volatility is illustrated in Figure 2.1. Energetic, if ill-informed, use of the industry as an economic regulator by successive governments, combined with speculative activity in the various markets for property meant that stop/go, boom and bust patterns in the economy at large tended to be greatly amplified in the construction sector.

Banks and stock markets are far less tolerant of construction risk and other forms of uncertainty than the founding proprietors were. The contractors were forced dramatically to reduce both the capital employed and the general risk levels of their businesses. As might be expected, this accelerated the flight of large firms from old-fashioned, self-perform general contracting. Many have attempted to shed their association with construction entirely, focusing on facilities management and aiming to become pure service firms, or at least acquiring that classification on the London Stock Exchange.

These processes do not seem to have seriously diminished the influence of the larger old firms in the industry, as Figure 2.2 shows. Most of the big players of 1986 – big since the war – are still recognisable, in one form or another, in today's league tables.[11] Their dominance has become increasingly precarious, however. Their territory is being impinged upon by smaller regional firms and by European competitors. And their status as big beasts in balance sheet terms is being challenged by the larger specialist contractors. Although the top firms still carry out a huge volume of business, for the most part, they simply procure and organise the work of large numbers of specialist trade contractors. It is the specialist companies who actually provide the materials, the work-force, and plant necessary to carry out the work. As Figure 2.4 below shows, the large contractors' profits are tiny, with their collective contracting margins averaging just 1.9 per cent over the past 20 years. This reflects the fact that they have done what they had to do: slashed capital employed and shed risk.

One way in which they have reduced the riskiness of their activities is by reducing the amount of lump sum work they do. The latest RICS/Davis Langdon 'Contracts in Use' survey shows that between 1985 and 2007 the use of lump sum forms of contract (excluding design and build) had fallen from 69.5 per cent to 31.4 per cent, and that management forms, including partnering agreements had increased from 17.1 per cent to 26.4 per cent, both measured by value of work. Management forms of contract are fee-based and supposedly riskless. In the same period, design and build, another supposedly low-risk form, increased from 8 per cent to 32.6 per cent.[12]

The withdrawal of the large contractors from the operational areas of the industry has created space and opportunities for smaller, nimbler, specialist firms to compete. While all this juggling of positions has been going on, the post-war period has seen a huge amount of innovation in the methods and materials used in building construction. Forty years ago, for example, the popular Barbour Index of building products comprised a set of ring binder folders occupying about six inches on a typical library shelf; today's equivalent is an on-line library of several gigabytes representing the offerings of over 7,500 manufacturers.[13] Many of these new products are genuinely new in the sense that they enable things to be done that could never have been done before. However, many new products are replacements for traditional components and materials. In some cases their novelty lies in their physical properties: lighter, stronger, higher U value and so on,

---

11 For the student of the structure of the UK construction industry, Hewes Associates' annual compilation of league tables, both contractors and consultants, for *Building Magazine* is fascinating material.
12 RICS, Davis Langdon, *Contracts in use. A Survey of Building Contracts in Use During 2007* London: RICS, 2009.
13 http://www.barbourproductsearch.info/index.html.

UK industry background   21

1987		2000		2009	
Company	Contracting £000s	Company	Contracting £000s	Company	Contracting £000s
Trafalgar House Plc	2,368,500	Bovis Lend Lease	3,931,034	Balfour Beatty	10,339,000
Tarmac Plc	2,163,100	Balfour Beatty*	2,504,000	Carillion	5,426,500
Wimpey (George) Plc	1,442,000	AMEC	2,475,300	Laing O'Rourke	4,087,100
Balfour Beatty Ltd	1,127,000	Skanska Construction	1,300,000	Morgan Sindall	2,213,500
Laing (John) Plc	1,063,800	Mowlem	1,153,000	Kier	2,145,600
Costain Plc	839,800	Carillion	1,129,400	Interserve	1,906,800
AMEC Plc	793,600	Kier Group	937,400	Newarthill	1,631,103
Mowlem (John) & Co	793,000	HBGC & E Nuttall*	929,300	Skanska Construction	1,541,800
Taylor Woodrow Plc	742,300	Laing	854,200	Galliford Try	1,510,300
McAlpine (Alfred) Plc	578,433	Newarthill	575,053	Bovis Lend Lease	1,444,995
Bovis Construction	382,291	Taylor Woodrow	500,900	Vinci	1,162,616
Newarthill Plc (est)	343,263	Mansell*	457,001	BAM Construct UK Ltd	1,133,600
Lovell (YJ) Holdings	323,754	Interserve	453,100	Costain	1,061,100
Higgs & Hill Plc	267,114	Birse Group	414,919	ISG	1,049,164
Crest Nicholson Plc	222,810	Wates Construction	400,758	Keller	1,037,900
Wates Building Group	209,276	Ballast	390,153	Willmot Dixon Group	1,003,413
Fitzpatrick Plc	208,191	Costain Group	386,300	Wates Group	945,250
Bryant Group	201,100	Norwest Holst*	381,767	Bowmer & Kirkland	873,591
Miller Group Ltd	190,493	Alfred McAlpine	374,075	Miller	783,000
Shepherd Building	186,666	Bowmer & Kirkland *	366,837	ROK	714,800
Boot (Henry) & Sons	153,367	N G Bailey	358,521	Shepherd Building Group	701,000
Wiltshier Plc	144,455	GallifordTry	355,427	Mace Group	653,613
Tilbury Douglas plc	140,184	Montpellier	339,171	BAM Nuttal	643,708
Biwater Ltd	136,300	Miller	318,117	N G Bailey	600,319
Galliford Plc	134,312	Morgan Sindall	317,605	Keepmoat Ltd	570,470
Croudace Ltd	98,400	Keller Group	312,954	Southern Electric	482,195
Gleeson (MJ) Group	91,759	Drake & Skull*	295,692	VolkerWessels UK	456,570
Carter (RG) Holdings	79,548	Willmot Dixon*	275,031	Biwater	412,200
Johnston Group Plc	79,478	Morrison	246,881	Severfield-Rowen	349,417
Longley Holdings	74,021	MJ Gleeson	241,115	Leadbitter Group	337,205
Birse Group Plc	65,950	McNicholas	236,926	Osborne	333,707
Mansell (R) Ltd (EST)	62,825	Henry Boot*	226,787	Byrne Ltd	326,221
Seddon Group Ltd	62,749	Shepherd Building	212,144	EMCOR Group UK	320,887
May Gurney Holdings	56,147	J Murphy & Sons	164,800	Renew	316,648
EBC Group Plc	52,675	R.G Carter Holdings*	160,149	R.G Carter Group	312,563
Willmot Dixon	52,477	Fitzpatrick	157,938	Spie Matthew Hall	312,331
Bloor Holdings Ltd	52,291	R O'Rourke	157,931	Barr Holdings - 9 months	309,136
Haymills Holdings Ltd	50,201	Simons Group	151,192	Canary Wharf Contractors	276,182
Try Group Plc	48,028	Staveley Industries	140,100	Apollo Property Services	261,635
GA Holdings Ltd	46,063	Osborne*	135,059	Imtech Technical Services	251,767
Osbourne (Geoffrey)	42,069	May Gurney Group*	130,110	Seddon	248,578
Morrison	37,308	Peterhouse	129,209	Midas Group	242,955
J Jarvis & Sons Plc	35,625	Severfield-Rowen	128,930	Higgins Group	224,884
		John Sisk	122,188	Thomas Vale Construction	213,639
		Sunley Turriff	116,077	William Hare Group	213,572
		Dean & Dyball	115,373	Bouygues UK	210,261
		Seddon Group*	113,556	RGCM	208,413
		McNicholas plc*	111,794	Carey Group	198,189
		Tolent Construction	111,096	GB Building Solutions	193,482
		Dew Pitchmastic	105,655	Lorne Stewart	183,410
		Llewellyn Mg'ment*	104,703	United House	183,000
		T Clarke	98,364	McNicholas Construction	181,587
		John Doyle Group*	93,900	Brookfield - 18 mths	178,904

Figure 2.2   UK construction – top firms, by contracting turnover

than their traditional counterparts. In many cases however, their attraction lies in their relative ease of installation.

One result of these developments has been a dramatic reduction in the craft component of on-site operations. Modern commercial and retail buildings particularly, involve a wide variety of specialist assembly and installation skills, but very little old-fashioned craftsmanship; one thinks of elements like curtain wall cladding, raised floors, suspended ceilings, internal partitioning and other systems-based components. This movement away from craft-based working, where much of the building fabric is created from raw materials, cut and shaped on site, will continue. Increasingly buildings will be made from kits of parts, manufactured and assembled off site and installed by site teams with high levels of narrowly specialist skills, but requiring little, if any, of the expertise of traditional craftsmanship. The very idea of craftsmanship – exquisite manual dexterity in the shaping of particular materials, acquired through long years of service – is disappearing from all but the most archaic (and expensive) areas of construction.

Second, although on the one hand trades individually are becoming de-skilled, on the other hand the large numbers of separate trades on a modern job site and the complexity of the interfaces between them require very high levels of skill to organise and coordinate. This is similar to the development of production-line techniques in manufacturing: the basic production processes were dramatically de-skilled, but the design and management of the production line itself became far more sophisticated, requiring high levels of skill which became concentrated in the hands and minds of production engineers and managers. Manufacturing industry shifted from being craft based and labour intensive, to being knowledge based and capital intensive. Logically, construction should follow that path.

The issues confronting the contracting side of the industry: specialisation/ fragmentation, innovation and selective de-skilling, are reflected, if anything more harshly, amongst the professions. Forty years ago the architect was almost a true *archi-tecton*, in the master builder sense. His role as the professional leader and manager of the project – uniquely knowledgeable, wise guardian of the public interest, and trustworthy beyond question – was highly cherished and fiercely defended. But, as new products, materials and construction techniques have been introduced into the industry, the architect's special body of knowledge – the counterpart of the craftsman's manual dexterity – has become less relevant, less valued. And following the dreadful mistakes made in rebuilding the centres of Britain's towns and cities after the war, the public seem to prefer lawmakers and bureaucrats to defend their interest through rules and regulation. Nobody trusts anybody these days; why should architects be different?

Well, so it must seem. The upshot is that today's architect is little more than a designer of buildings, the first amongst equals in the new project team perhaps, but no longer the dominant, unifying force that the traditional idea of the profession represented.

It's a strange coincidence that, over more or less the same time period, and in much the same way, both of these major players in construction – the main contractor and the architect – should have vacated the space they have traditionally occupied at the centre of projects. The vacuum left behind has been filled by a variety of new people; who knows how things will shake out over the longer term.

Morton and Ross[14] tell this story in more detail and place recent developments in the longer historical context; and Morledge et al.[15] provide a useful review of the structural changes and related contractual innovations of the past 50 years or so.

This section has discussed four key features of the UK construction industry:

- the idea of project teams as being the true operational focus of decision making and innovation in the industry;
- the industry's long tradition and active pursuit of improvement;
- its relatively malleable organisational structure; and
- its under-appreciated, but very significant capacity for basic innovation.

These all contribute to a background that lends itself readily to the adoption of methods like Building Information Modelling. They will be re-visited in the context of the BIM futures analysis in Chapter 9. The second half of this chapter considers the reasons why BIM is necessary.

2.2 Strategic challenges

Arguably, the defining characteristics of the modern construction industry are:

- its inability to complete projects predictably;
- its chronically low level of profitability.

Predictability is essentially a matter of completing projects on time, on budget, and to the agreed level of quality. Constructing Excellence (CE) record that small improvements in certain other areas of the industry's performance have been achieved over the past ten years. However, as Figure 2.3 shows, no such improvement has been achieved in the area of cost and schedule predictability.

Despite all the effort and exhortation of the past decade, more than half of all UK construction projects exceed either their contract budgets or agreed schedules, or both. The CE programme has demonstrated that this is true regardless of project size or mode of procurement. The headline figures shown in Figure 2.3 are extracted from CE's most recent report on the industry's performance. What this chart doesn't show is that half of these projects – that is, a quarter of all UK construction projects – overrun budgets or schedules by 10 per cent or more. This sort of outcome can have devastating effects on businesses and on the careers of individual people who work for them.

---

14 Morton R. and Ross A., *Construction UK*, 2nd edition. Oxford: Blackwell, 2008.
15 Morledge R., Smith A., and Kashhiwagi D.T., *Building Procurement*. Oxford: Blackwell, 2006.

24 UK industry background

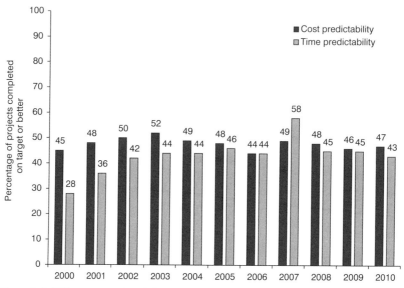

Figure 2.3 UK project predictability, 2000–2010

Unpredictability has two main adverse effects. First, schedule overruns have the direct effect of increasing overheads and preliminary costs. These may be recoverable, but usually only if they are pursued as claims with the client. This process is generally disruptive of good relations and can be downright confrontational. However they are handled, schedule overruns usually result in direct loss of profit to the contractor.

However, the main adverse outcome of project unpredictability is disappointed, often distressed clients. The effect of this is to make those clients – the firms, as well as the individuals concerned – reluctant to repeat the experience and generally averse to dealing with the industry. No data exist to support this suggestion, but it would seem reasonable to suppose that on aggregate, across the economy, the overall level of demand for construction services is suppressed by at least one or two per cent, as potential clients seek alternative ways of solving their accommodation problems.

The combined result is that industry profits are hit twice: first, by the direct impairment of margin caused by overrunning projects, and second, by the loss of business volume caused by disaffected clients seeking alternatives to construction.

Serious overshooting of project schedule and cost targets can be disastrous for the individuals and firms involved. But it's conceivable that the situation might not be a major cause for concern for companies if their losses on failing projects were offset by substantial profits on other jobs. Unfortunately, they're not. Hewes Associates – compilers of *Building Magazine*'s construction league tables – carry out an annual assessment of the margins (profit on contracting turnover) of the top firms in the industry. Figure 2.4 plots the results for the past 20 years. It paints

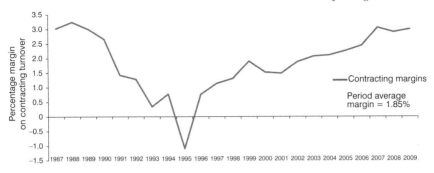

Figure 2.4 Major contractors' contracting margins 1987–2009

a pretty dismal picture. An average margin of 1.85 per cent leaves no room for error and, depressingly, little scope for investment.

Three supplementary points are worth noting. First, as discussed earlier, few of the main contractors in the UK industry self-perform any significant proportion of the work on their projects; they use specialist sub-contractors almost entirely. They don't therefore employ any substantial amounts of fixed capital on their projects. So any positive margin might be said to represent a high level of return on capital – the basis on which most investment decisions are made. The reality is that the main contractor is usually the one with overall liability for delivery of the project; any failing redounds on him. So, as Laing Construction found in 2001, the risk-weighted return is a great deal less than the headline number.

The second, related, point to note is that it is almost impossible for players in this market to maintain effective barriers to entry. For main contractors, capital, in the traditional form of construction plant and equipment, is no longer required, so sheer size of balance sheet is no longer an effective defence. And their low levels of profitability make investment in people, or in intellectual property through R&D, effectively impossible. So even knowledge-based barriers to entry are not available; as recent extinctions have shown, the very survival of even the largest, ostensibly most successful UK contractors, is a brutal struggle.

The third point is that a large part of the income of the main contractors derives from the timing and management of the flow of client payments into their bank accounts and subsequent payments out to suppliers and sub-contractors. This is estimated to be about two per cent of turnover – roughly equal to their construction margins. It's little wonder therefore that payment disputes are so prevalent in the industry and that 'pay-when-paid' remains such a bone of contention.

2.3 Solutions

It may seem arbitrary to focus on just two of the many problems confronting the UK construction industry at the present time. Why predictability and profitability particularly? Why not sustainability, productivity, collaboration, safety, skills, or a host of other, generally more topical problems? The answer is simple. Although these are all important topics, they are actually second order issues. They are not fundamental threats to the survival of construction firms; predictability and profitability are.

And that's why BIM is so important. Unpredictability and low profitability are both caused in large part by the same underlying phenomenon: the devastatingly poor quality of most of the information used on modern construction projects. BIM targets that problem precisely.

Construction is an extraordinarily communications intensive industry. A European construction IT R&D project called CICC,[16] in 1995, found that up to 400 individual documents, or documents about documents, are generated for every million pounds' worth of project value. CICC also found that there may be up to 60 consulting and contracting firms involved in a typical £50m project. The large numbers of people, the huge numbers of documents and the speed of circulation, all add up to an environment of exceptional information intensity. The crucial point is that very little of the information currently generated in construction is structured, systematic or trustworthy.

Typical construction documents – drawings, instructions, schedules, programmes, bills, certificates, reports and so on – in fact comprise more or less shapeless masses of ambiguous, subjective information, largely lacking in systematic content. To interpret and use this material accurately and consistently requires the application of very high levels of human judgement and intuition; skills that are both rare and largely un-teachable.

Broadly speaking, business communications can be broken down into three main elements:

- the *context* in which the communication is taking place – in the present case, specifically, the commercial and contractual relations between the communicators;
- the *communicators*: the firms and individual people actually doing the communicating;
- the *content*: the nature and quality of the information being communicated.

Most of the post-war studies referred to in Section 2.1.1 were concerned with the first of these issues. They focused their analysis and recommendations on the forms of contract and other aspects of the commercial framework in which the industry operates; to very little lasting effect, as Sir Michael Latham notes at Executive Summary point 9 in his report.[17]

The Latham report in its turn focused mainly on the second of the components of communications; the communicators: the people, their attitudes, and the supposedly adversarial culture of the industry. Latham and his successor Sir John Egan promoted partnering, frameworks and other forms of collaborative ways of working. The idea was that by building teams which put aside the apparently

---

16 CICC, EC ACTS R&D Project 017, 1995–97. Brussels: EC, 1998..
17 Latham, Sir Michael, *Constructing the Team*. London: HMSO, 1994.

innate mistrust and antagonism that pervades traditional contractual relations, one might improve project performance. Confrontational attitudes *per se* are seen as an important cause of project failure. The Latham/Egan hypothesis might be said to be: 'Reduce/eliminate confrontation and you will reduce/eliminate project failure.'

However, as the Constructing Excellence data show (see Figure 2.3 and most of the other key performance indicatores (KPIs)) despite over ten years of sustained high-level exhortation and a huge investment of effort and goodwill on the part of individual project teams, no sustainable or perceptible improvement has been observed either in project predictability or in company profitability – the most crucial of CE's key performance indicators. Sir John Egan himself wrote on the tenth anniversary of his original report: 'I'd probably only give the industry about four out of 10, and that's basically for trying, having its demonstration projects, still being in the game, and still having enough there to actually, perhaps with another big heave, get it done the next time around.'[18]

A more recent review of progress, led by Andrew Wolstenholme, a former colleague of Sir John's at BAA, was carried out in 2009.[19] Wolstenholme reported largely the same lack of achievement as Sir John had done, but seemed to suggest that more of the same – something like his predecessor's 'another big heave' – would bring the breakthrough. The report proposed a 'new vision for the industry' and exhorted 'suppliers, clients and government to think *built environment*'. However, the lack of a convincing explanation of the causes of the industry's fundamental problems means that the report's recommendations add little to Sir John's exhortation to the industry to try harder, do better. The proposition that the current generation of industry leaders should be prepared to up their game or 'step aside and let others take over'[20] is, frankly, unlikely.

Even if the future construction industry envisioned in the Latham/Egan analysis were fully convincing, there is no mechanism proposed by which it might come about, or even be caused to happen – apart from the exhortation of Sir John, politicians and others. Exhortation and persuasion may work for a limited period, with a limited number of sympathetic listeners. But to attempt just by talking to it, to change the trajectory of something as massive and insensate as an entire industry – let alone one as intractable as construction – is not really likely to work. The only proven incentives to change are regulation and profit. Construction is arguably over-regulated already, and Latham/Egan has shown no convincing link between collaborative behaviour and profitability.

So, focusing on contracts didn't work, and focusing on people hasn't worked. It's time to attend to the third component of communications in the industry: the

---

18 Egan, Sir John, 'I'd Give Construction About Four Out of 10', *Building Magazine*, 8 May 2008.
19 Wolstenholme, A., *Never Waste a Good Crisis: A Review of Progress since Rethinking Construction and Thoughts on Our Future*. London: Constructing Excellence, 2009, p 4.
20 Ibid., p.25.

nature and quality of the information used in construction. Higgin and Jessop,[21] in one of the reports discussed in Murray and Langford above, actually did just that. With its focus on organisational structures and information exchanges, Higgin and Jessop is one of the most insightful and interesting of the historical reviews, particularly in the present context. (Unfortunately it seems that organisational politics intervened; Higgin and Jessop's initial pilot work was never properly followed up.)

The present hypothesis suggests that the direction of causation should be the reverse of that described by Latham/Egan: projects don't fail because people are defensive and confrontational; people are defensive and confrontational because projects fail. The underlying reason why projects fail is because the industry persists in trying to build highly complicated things using incredibly poor-quality information. This information is largely untrustworthy, in the sense that it cannot generally be taken as being : 'factual, without need for evidence or investigation' (*New Oxford Dictionary of English*). As every trainee site engineer knows, everything you read in construction must be checked, and then checked again.

'The management of construction projects is a problem in information, or rather, a problem in the lack of information required for decision-making.'[22] The very first sentence in Graham Winch's book says it all. Contracts, claims and litigation are all part of a system that helps firms to cope with this situation. Without them no building would ever get built. The danger with 'collaborative' forms of contract is that they entice firms into putting aside these protections. What looks like a bridge over the swamp often turns out to be just a diving board. They glide over the problem of information quality. They say 'Trust me, I'm your partner, I'm your friend, I won't do badly by you'. They don't deliberately abuse each other – it's the information that does the damage. And it's an unfortunate fact that, when firms and individuals place their faith in an idea like personal trust and collaboration, and find that that idea comes undone, the result is often more bitter than even the most rancorous conventional dispute.

Before truly collaborative, trust-based forms of relations can be developed in construction the industry must have truly high-quality information – material that can be trusted, that can be taken as true, without need for evidence or investigation. That is the right order of things and that's the really important thing that BIM offers: dramatically higher quality information and communications processes. This development feeds through directly and logically into dramatic improvement in predictability and profitability as will be demonstrated in the next two chapters.

There are several other areas of modern life where, by radically improving the quality of the information they use and the means by which they deploy it, individual firms or even entire economic sectors have dramatically improved their performance and their customer satisfaction. Banking with ATM; airlines with yield management; manufacturing with CAD/CAM are a few that come readily to

21 Higgin, G. and Jessop, N., *Communications in the Building Industry*. London: Tavistock Publications, London, first published 1965, reprinted 2001, Routledge, London.
22 Winch, G.M., *Managing Construction Projects*. Chichester: Wiley-Blackwell, 2010, p. xiii.

mind. But arguably the most notable example of this has been in the transformation of the retail sector following the deployment of point of sale (POS) systems, starting about 30 years ago. POS is a retail data management tool. An electronic point of sale system is essentially an information model of the store in which it is being operated. The system captures enormous numbers of tiny pieces of very precisely specified data about the state of operations in the store, as purchased items of stock flow through its checkouts. The system aggregates and organises this data and provides a variety of analysis and management tools, which can be used by store managers, buyers and others to manage the business of the company. Modern POS systems give retailers almost complete control over both their own operations and those of their entire supply chains. In place of guesswork, intuition and subjective judgement, POS provides far more reliable, data-based, technical methods of status assessment and stock control – the essential requirement for efficient, large-scale retail operations.

The central thrust of the present argument is that both of the fundamental problems of construction – lack of project predictability and low industry profitability – result directly from the industry's excessive use of low-quality, unstructured information and the degree of dependence on human judgement that this necessitates. The only way to make projects predictable and firms profitable is by substituting computable data for unstructured information. Building Information Modelling (BIM) and related tools and techniques will enable this to happen and will transform construction, in much the same way that POS has transformed the retail sector over the past 30 years.

3 The problem

3.0 Introduction

So far our discussion of the problems caused by poor information and inadequate communications in construction has been expressed in fairly high-level terms. This and the next chapter take the discussion down to the operational level of the individual construction project. At this level of detail the issues raised take on a more tangible aspect, which provides a more useful basis for devising specific, workable, corrective actions.

A model-based design process generates far higher quality information than is possible with drawings, and the standards-based interchange of the resulting computable data dramatically improves communications amongst the project team. Both of these developments have profound consequences at the operational level where the key potential benefits of the BIM approach become clear.

3.1 General features of drawing-based design information

Most of the information about a given project originates in the architectural design, primarily in the form of architectural drawings. Using conventional techniques, a designer works by creating pictures of the object that he has in mind. The pictures are made by drawing lines to represent the edges of objects, or of details within objects, either on paper or on a computer screen. The lines can take the form of simple straight lines, polygons, curves, ellipses, circles, arcs and so on. Each one of these lines constitutes an individual, discrete piece of information, with its own origin or point of insertion into the drawing, as well as its own length, direction and other geometric characteristics. Drawings may be enhanced using a variety of codified line styles, hatch patterns, colours and the like.

Technical drawings rely heavily on the use of discipline-specific symbol conventions and annotations to convey information in an efficient manner. Such annotations are generally separate as drawn entities from the objects in the drawing to which they refer. For example, the text denoting the length of a line is an entity separate from the line itself. A draftsman wishing to change the line length must also change the text, as a separate task, in order to keep the two in synch.

In order to represent the object adequately the designer may need to provide a number of different views. Thus for example a simple three-dimensional object usually requires at least three separate views to represent it adequately: a plan view, from directly above; a section through the object on a representative plane; and an elevation, or side view, of the object in question. Every drawing represents a separate, individual view of the thing being designed, and every view of the object needed to represent it definitively requires the preparation of a separate drawing. The result is a enormous number of individual lines and annotations, which must be managed across all of these individual drawings. And of course there are also many other forms of information, such as specifications, construction instructions and schedules that must be created and associated with the relevant objects in a given set of drawings.

The difficulty of maintaining consistency across a single set of documents of this sort is obvious. A compounding level of complexity is introduced when such a package of information, from the architectural practice for example, is passed to others such as the consulting engineers for them to use as the basis for their own sets of lines on their own sets of drawings, and other documents.

There follows the problem of coordinating all of these documents – which must in many respects be done at the level of detail of the individual drawn line – through review and approval processes of varying degrees of complexity, before they can be passed onwards to the next link in the supply chain. Considered objectively, it hardly seems possible, but this is basically how even the most sophisticated of modern buildings are designed.

All of these aspects of manual drafting practices were, necessarily, embodied in the development of early computer-aided drafting (CAD) systems. And though many improvements have been made over the years, it is generally true that drawing-based CAD systems of this sort continue to suffer from the same basic limitations that beset manual design techniques:

- The inherently cryptic nature of conventional design documents. The reader of a drawing must make assumptions and judgements about any part or detail of the thing being designed that is not shown explicitly on the drawing. This leads to many types of misinterpretation. At a basic level, someone who is not familiar with the forms of notation and symbology being used will be prone to misunderstanding the meaning of the drawing.
- The difficulty of coordinating between related drawings to ensure that a detail that appears in more than one view of an object is properly represented in all the corresponding drawings.
- The difficulty of associating data about the thing being designed with its representation on the design drawings. The lines that make up a conventional drawing are just that; lines on a piece of paper or on a computer monitor. They contain no information, and no significant information can be attached to them.
- The difficulty of representing complex shapes and forms. Complicated shapes, particularly objects that change section in more than one plane

simultaneously, can actually be impossible to represent using conventional drawings or CAD images.

More recent CAD systems help to manage these problems. For example they enable individual lines to be grouped into polylines, cells or blocks and they offer basic data management capabilities. Nonetheless, the fact remains that most modern building designs are created using basic line primitives laid out in two-dimensional space.

Thus, this conventional drawing-based design process generates inherently low-quality, unstructured information that it is essentially un-trustworthy. By this is meant that anyone who receives such information cannot assume that what it contains is true. Instead, before it can be used, it must be checked to ensure that it is:

- correct, as far as the recipient is concerned;
- clear and unambiguous in its content and representation;
- consistent internally and with other documents produced by the same author or firm;
- properly coordinated with related documents produced by others;
- complete and sufficiently detailed for the recipient's purposes.

To carry out these checks effectively and consistently takes time and requires extraordinarily high, but generally unacknowledged, levels of skill, discipline and judgement. Such talents are rare and often unavailable on fast-moving projects, which means that fundamental mistakes are often made.

Also, the lack of structure in this sort of information means that, even though it may well have originated in a computer system, it cannot be reused directly in the recipient's computer. It must first be decoded or interpreted, then re-keyed or otherwise re-entered into the recipient's system.

These two facts account for much of the under-performance and dysfunctional behaviour characteristic of today's construction industry. More seriously (if that's possible), these problems of information quality in construction, if left unresolved, will lock the industry for the foreseeable future in its current, essentially pre-industrial, craft-based mode of production.

3.2 The impact of poor information on design processes

Figure 3.1 illustrates this discussion in a highly abbreviated form. In this model of the project process, the overall project is broken down into its three main phases: design, procurement and construction. The principal types of information used in each phase are indicated. The commonest flaws in the available information are highlighted, and the consequences of its use are suggested. These consequences are expressed mainly in terms of the two strategic industry problems discussed in Section 2.2: predictability of projects and profitability of firms.

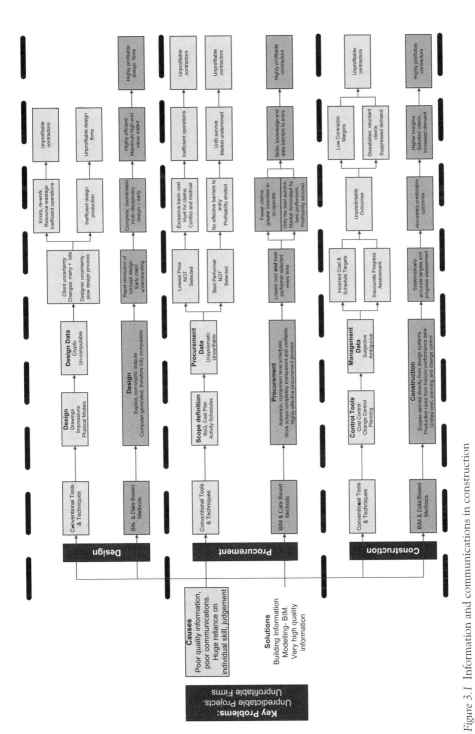

Figure 3.1 Information and communications in construction

For the purposes of this discussion, building design can be broken down into two broadly separate areas of activity: actually doing the design work, and communicating the results of that work to other individuals and firms involved in the project.

3.2.1 Problems with drawings in design production and administration

The main problem pertaining directly to the design activities of a single practice is ensuring that the information generated within the practice is complete and consistent. Design team leaders and others must spend a large proportion of their time checking their teams' work, ensuring that all the necessary documentation has been produced and that it is of the required standard of content and presentation. Team leaders must also ensure that every view or description of any given object in any given document is complete and that no unspecified geometrical or functional gaps or overlaps occur. This is tedious and difficult work, for many designers more arduous than the actual creative design activity – a huge expenditure of high-value effort, unrecognised and so, poorly rewarded.

There are many other, lower value administrative chores that drawing-based design processes make necessary: maintaining CAD standards and filing systems, maintaining document registers, administering change control systems, evaluating and administering requests for information and change proposals and dealing with responses, instructions and so on.

These are just a few of the low-value chores that consume the time and attention of design teams. Modern, pre-BIM CAD systems and related applications can assist with some of these tasks, however, these systems can be double-edged in this respect. Like many other types of computer-generated documents, CAD drawings can acquire a credibility that is sometimes unwarranted. People trust them more than they should and don't check them as well as perhaps they should. A second problem related to the use of conventional CAD systems is that, because CAD drawings are so easy to produce, more of them are sometimes created than are strictly necessary. This obviously exacerbates the checking problem, without necessarily improving the overall quality of the design information.

In these and other ways, dependence on drawings imposes a significant administrative burden on all of the members of the design team. It's notoriously difficult for firms to obtain adequate payment even for the clerical component of this work, let alone for the time and effort required of senior designers in the administration and checking of drawing-based information. The cumulative effect on the individual design practice of its dependence on drawings as the basis of its design work is a direct loss of profit to the firm, and a crushingly tedious waste of the time of talented designers.

So, dependence on drawings as the basis of the design process leads to cumbersome, tedious, low-value administration and checking work within the individual design practice. It also leads to problems for all of the other members of the greater project team.

3.2.2 Problems in communications with the client

Problems arise when communication requires judgement or interpretation on the part of the recipient of any given piece of information. In the current context, there are three main groups of recipients: the client and his stakeholders, the other members of the design team, and the construction contractors. Problems of interpretation using drawings and conventional design communications tools and techniques impact differently on all three of these.

The client and his team are usually lay people for whom architectural and technical drawings are more or less incomprehensible documents, both in terms of the architectural details they represent, and in terms of the spatial arrangements they are intended to convey. Architectural illustrations, physical models of wood and plastic and such like, even the early forms of computer-generated impressions and 'walk throughs' are of relatively little value.

The lay viewer remains unsure, sees something new in every iteration of the design and reacts with a continuous stream of queries, requests for change and such like. The client's uncertainty introduces delay and revisions which reverberate throughout the entire design effort. The efficiency of the design process is undermined and the potential profitability of all of the participating firms is impaired.

3.2.3 Problems in communications with the design team

Problems of accurate communication and understanding also arise in regard to the architect's supporting disciplines. When the design output takes the form of paper drawings, which have to be inspected and coordinated by eye at every issue and every revision, these problems can be particularly severe. Given the cryptic, discipline-specific language of technical drawings, errors of understanding are almost unavoidable. Much effort is wasted and profitability is undermined.

Even when the design firms exchange or share editable CAD files with each other, problems of interpretation and integration arise. It is surprisingly common for simple coordination errors to be overlooked, and such unlikely mistakes as scaling errors, mismatched origins, misaligned grids and the like occur disturbingly frequently. These problems impact heavily on the efficiency of the design process, introducing errors and the need for re-work, again reducing efficiency and impairing the profitability of the overall design effort.

3.2.4 Problems in communications with the contractors

The third area where problems of design communications arise is at the interface with the construction team. There are two main types of difficulty at this point. First, the inefficiencies and time lags introduced by the problems outlined above cause disruption and delay to the design production programme with a variety of adverse impacts on the subsequent procurement and construction phases.

The second problem in this area is that, like the blind men and the elephant, everyone who looks at a set of technical drawings sees something different. The differences may be slight and easily reconciled, but frequently even experienced construction people misinterpret what they think they see. This problem is particularly acute when multiple orthogonal projections are required to create an understanding of a complex three-dimensional object or space. The problem is greatly compounded when the thing to be visualised comprises multi-disciplinary elements.

Delays and disruption in the design delivery programme and errors in interpretation and understanding of the content of design documents all lead to drastic inefficiencies in the basic operations of the industry. More importantly perhaps, they make participants hesitant, uncertain, and defensive in their dealings with each other, which makes them generally reluctant to trust and therefore to innovate.

3.3 The impact of poor information on the procurement process

Price competition is the driving force of the market economy. It ensures that buyers get the goods and services they require at the best possible combination of price and quality. In order to survive in this competition, suppliers are forced to innovate continuously, either to supply at the lowest price or to supply the highest quality. Suppliers who are unable to compete are eliminated from the market; only the best survive. Thus, effective competition has two hugely beneficial effects for the economy: goods and services supplied at the lowest prices, and continuous improvement of products and services by suppliers

The lump sum, fixed price competitive tender is, in theory, the closest approach that exists to pure price competition in the selection of construction contractors. All other things being equal, this should be the best mechanism for arriving at the lowest cost outcome on individual contracts. It should also be the most reliable way to embed effective competition – and thereby real, sustained improvement – in the industry at large. As will become clear, competition in construction today – intense thought it may appear to be – achieves neither of these goals.

The difficulty is that, for price competition to work effectively, the buyer must be able to specify his requirements fully, accurately and unambiguously, and must subsequently be able to compare potential suppliers' competing offerings on an exactly like-for-like basis. The documentation used in construction procurement does not enable this to happen.

3.3.1 Bills of quantities

The most systematic and, in theory, the most precise form of procurement documentation used in construction is the bill of quantities (BQ). There are many other forms of procurement documentation, generally less rigorous than the BQ, which all suffer even more than the BQ approach from the problems outlined here. So for present purposes, the BQ is taken as the exemplar.

Although it may be used for other purposes in the course of the project, the central function of the BQ is to provide an accurate specification of the scope of work of the contract at hand, delineated in such a way as to enable accurate, like-for-like comparisons of the bidders' proposals. In the standard BQ preparation process the taker-off locates, classifies and measures each of the individual building components, as he observes them on the relevant drawings. He then groups the details of these individual items into class aggregates, according to the rules of the relevant Standard Method of Measurement (SMM). For example if the project contains 120 reinforced concrete columns of varying cross-sections and heights, at different locations throughout the building, all 120 might well be described in three clauses: one for the formwork, one for the reinforcement and one for the concrete – 'Concrete, class A, in columns – 928 m^3', for example.

Such drastic compression of the original detail is necessary because paper is the traditional medium of communication of BQs. Without this level of compression the project might have to deal with the storage, copying and distribution of thousands of pages of bills – a huge and cumbersome job. The computational power and storage capacity of modern computing systems means that none of this is any longer actually necessary. But the tradition persists. The result is that although the take-off surveyor may record a great deal of potentially useful information about each of the individual components of the building, he must then effectively throw away almost all of that material in order to produce the sort of terse, cryptic, SMM-compliant line items of which the example above is typical.

The fundamental problem with the BQ approach is that the compression process that's used in the creation of the bill is not reversible. It is not possible to map line items in the bill unambiguously to elements or components of the building as shown in the drawings. So, for a given bill item, even if one adheres strictly to the rules of the SMM, it's impossible to re-generate the component-level detail that gave rise to it. It is therefore impossible to ascertain the true scope of a particular line item, which in turn makes it impossible to know the true scope of the contract as a whole.

The contractors bidding for a project know that once they sign the contract, they are signing up to the scope as supposedly specified in the bill. In order to ensure that he has covered the scope fully, each of the bidders for a particular contract must therefore carry out his own take off and attempt to reconcile this with the buyer's bill. Given today's industry of layered contracts, sub-contracts, sub-sub-contracts and so on, the resultant duplication of tendering effort involved in procuring even the simplest building is astonishingly wasteful.

A second problem with the BQ approach is that, for the reasons outlined above, even the most diligently prepared bill contains a large proportion of line items that cannot be verified precisely. Each of the bidding firms interprets the bill in its own way and responds in its own way, based on this interpretation. This makes it impossible for the buyer to arrive at a set of definitive, accurately comparable responses. Instead, the procuring party, with each bidder, must carry out a variety of 'normalisation' or 'bid conditioning' exercises – adding a little scope and cost here, deducting a little there – massaging the numbers until he's got a proposal that he and the bidder can agree on. This will hardly ever be the real, lowest possible bid for the work, but no-one will ever be able to discover that, either to prove or to disprove it.

A third problem with loosely or ambiguously defined work scopes, as represented in bills of quantities, is that all the bidding contractors must factor in scope risk in their tenders. Regardless of whether any of the bidders is behaving in a predatory manner – win at all costs, then make your money in claims – each of the contractors, if he wants the work, must assume that someone on the bid list will bid unreasonably low, either inadvertently or deliberately, therefore he must do the same. He cuts his core unit rates to the minimum, cuts his allowance for overheads and preliminaries, and cuts his margin. Everyone must do this.

3.3.2 Dysfunctional competition

The result is that incompetent bidders, who fail to see what's involved, bid low, get the work and lose money. Predatory firms who see the claims opportunities, bid low, get the work and make their profit from claims. Competent, non-predatory firms, who make reasonable allowances in their bids, bid higher and lose the job, or bid lower and lose money. That, in a nutshell, is how competition works in construction today. That process, more than any other single factor, is responsible for the high risk/low profit combination that characterises firms in the industry.

The problem is that the bidders effectively end up competing for the claims opportunities inherent in poorly documented designs, because there is simply no profit to be made in the core scope of work. As a result, competition in the industry is not between the construction competence of rival firms, but between the capabilities of their respective estimating and claims departments.

The huge waste of effort and the generalised failure of conventional procurement techniques to achieve the lowest price from project to project arguably has little impact beyond the projects in question. However, the failure of the process to ensure the selection of the 'best' bidder is a far more pernicious problem for the industry over the longer term. Best bidder in this context is taken to mean the firm most capable of performing the construction operations required. Less competent firms, with better estimating or claims capabilities are able to survive, as predators, in this system. These are not the firms who would survive in an efficient market. But this is not an efficient market, functioning as markets should. It's a dysfunctional travesty giving rise to a profitless, subsistence industry, with no capacity for investment in either physical or human capital.

A construction industry with contract procurement based on Building Information Modelling would be an entirely different proposition.

3.4 The impact of poor information on construction management

In the construction phase, the problem of poor-quality information arises mainly in regard to project management activities: planning and scheduling, cost management, change management and related functions. Of course, the general problems of interpretation and judgement outlined at the outset permeate construction operations, but poor information in project management is the cause of the industry's strategic, predictability problem. (Peter Morris' excellent book on the management of today's projects elaborates on the systems aspects of this problem – the focus here is the information used in those systems.)[1]

Projects over-run budget and schedule targets for two broad reasons:

- either the targets were set incorrectly in the first place;
- or, inaccurate progress assessments are generated in the course of the work, leading to misread trends and inappropriate corrective actions, or both.

We will focus on schedule targets for the moment. In general, the initial target for the time required to complete a piece of work is a function of the total planned output, divided by the relevant historical rate of production. And, similarly, the interim forecast time to complete is a function of the total planned output, minus output completed to date, divided by the rate of production achieved on the project to date. Thus there are three key variables:

- total planned output
- output completed to date
- historical, or current average rate of production.

The problem in construction is that none of these variables is ever known precisely. Instead, approximations or even proxies are used in most contexts. To appreciate this more fully consider some of the key problems, in this particular regard, that apply in the conventional approaches to planning and scheduling.

The conventional planning function involves the creation of activity-based project models, generally in the form of CPM networks and Gantt charts. These models have two main purposes: initially to demonstrate that the project is physically/logically achievable; and subsequently, during the course of the work, to provide short-term guidance to action, in the form of look-ahead schedules and such like. But activity models are flawed in both of these areas of use, as the following will make clear.

3.4.1 Dependence on individual subjective judgement

Activity planning models are highly subjective creations. On a given project the particular activities chosen, the way in which these activities are defined

---
1 Morris, P.W.G., *The Management of Projects*. London: Thomas Telford, 1997, pp. 213 ff

or specified and the logic links used to connect them are all determined by the personal experience and intuition of the individual planner or project manager. There is nothing standardised or systematic about the activities. In a sense, an activity means what the planner wants it to mean. This has three highly undesirable consequences:

- First, the sense or scope of a given activity tends to be cryptic, difficult for other people to comprehend and therefore difficult to challenge or verify independently. This undermines the model's value as a test of the viability of the project. It also has adverse consequences for the model's subsequent use as a management control tool.
- Second, the meaning of the activity is not determined and fixed. It can be changed by the planner through the duration of the project. This means that analyses and supposedly like-for-like comparisons between different points in time on the project can be misleading.
- Third, activities tend to be very project specific, reflecting the individual planner's personal response to the project at hand. This means that there is little scope for cross-project comparison or analysis which, again, makes it very difficult for people other than the planner and his project manager to challenge the plan.

Also, as a result of the unsystematic way in which they are used to represent the scope of work of the project, it can often be difficult to carry out progress assessments on a consistent, like-for-like basis using activity models. There is often a tendency towards optimism bias and other assessment errors which tend to obscure the true situation. This makes performance trends difficult to spot and effective responses hard to devise.

3.4.2 Planning versus forecasting

It's important to distinguish between plans and forecasts. Once a given project gets under way, the main requirement of the plan is to provide guidance to action: rolling wave look-ahead programmes, short-term activity schedules and suchlike. To do this effectively, the plan must reflect accurately the current, real-world situation. It must therefore be revised more or less continuously to embody the effects of new information and changing circumstances. The only way this can be done is by re-defining, or re-specifying the activities that make up the model, or by re-arranging the logic links between activities.

Once this starts to happen, forecasting – like-for-like comparison of the model at different points in time – becomes impossible, and the ability to generate consistent long-term forecasts is lost. For this reason in particular, planning models make inherently poor forecasting tools. The planning model cannot, logically, be both a short-term guide to action and a reliable tool for consistent longer term forecasting.

3.4.3 The top-down problem

Conventional project management systems are essentially top down in their structure and operation. It is extremely difficult to integrate them closely with the detail of operations at the production level of the project. This places a great deal of reliance on the front-line supervisor to act as the interface, the interpreter, between the planning system and operations on the ground. For the most part individual supervisors can be relied upon reasonably well to understand the plan and to translate its contents into detailed task schedules.

However, when assessing and reporting progress achieved, even the most competent and most experienced supervisors are sometimes betrayed by what academics call optimism bias. This has been defined as 'a cognitive predisposition found with most people to judge future events in a more positive light than is warranted by actual experience'.[2] In the present context it refers to the intuitive reluctance of people to convey bad news to their superiors; instead they delay acknowledging poor progress, often until it's too late to correct. This reporting problem applies to all project management disciplines and is evident at all management levels in project organisations. Everyone involved interprets the information he's given subjectively – because there is no systematic, objective content – and adds his own twist to the story he receives, before passing it on. Sometimes the twist is slight, sometimes it can be critical. Insofar as it substitutes opinion for fact, it's always undesirable.

3.4.4 Lack of responsiveness

Projects very rarely fail catastrophically, completely without warning. There is almost always a history, some process or sequence of developments that leads up to an apparently sudden crisis. The problem is that planning systems are simply not designed to capture the sort of historical information that might be useful in detecting these sequences or trends.

They capture reasonably well the 'instant in time' snapshot pictures, cross-sections through the history of the project that are necessary for short-term activity planning. But these systems are very poor at knitting these cross-sections together or in other ways producing longitudinal, time-based views of the project's evolution. Planning systems simply don't cope with trends very well. So project teams often don't realise they are off target until too late and it becomes costly and disruptive to institute corrective actions, 'crash' recovery programmes and suchlike.

3.4.5 Inability of companies to learn

The four problems outlined above relate to the difficulty of using planning systems to predict project cost and schedule outcomes, while the project is under way. The

---

2 Flyvbjerg, Bent, 'From Nobel Prize to Project Management: Getting Risks Right', *Project Management Journal*, August 2006.

fifth problem is more to do with accurately establishing those targets in the first place. It relates to the way in which firms gather and use actual performance data from their projects.

The individual people who work on projects learn a great deal from every job they do; the companies they work for learn almost nothing. The human learning is, for the most part, experiential and unstructured. Companies are not able to 'learn' in that sense. Companies learn by gathering structured data that can be analysed, stored, evaluated and reused in future activities. In order to capture this material the data has to be specified systematically and there must exist organising frameworks or other mechanisms that enable it to be gathered and arranged efficiently for future use. CPM and similar planning systems meet neither of these conditions: the data used are too subjective and the activity models are too job-specific to be systematically useful on future projects.

3.4.6 Project management issues – recap and summary

The issues above have been elaborated upon because they are rarely, if ever, drawn out in discussions of project failure – despite the fact that unless the problems they represent are solved, other attempts to improve construction project predictability will almost certainly fail.

To summarise, our theory suggests that projects fail because conventional project management methods and systems:

- depend too much on intuitive, subjective definition of work scope and progress assessment;
- are dangerously top-down in their operation, lacking systematic connection with the production level in projects;
- are inherently poor for forecasting and for trend detection and analysis;
- provide no effective frameworks or methods for the capture, analysis and reuse of performance data.

Similar problems occur in the area of cost control. Again, the principal problem is that the project scope is not specified systematically or in useful detail. Cost models based on cost planning techniques – elemental costs per square metre for example – suffer from the same inherent problem of subjective definition as activity models, with the same results.

Bills of quantities (BQ) at least have the merit of being based on the detail of more or less complete design documentation. However, as discussed above, in order to make a BQ manageable as a paper document, most of the useful information recorded in its development has to be compressed out of the final product. Hundreds of lines of valuable, specific data about individual concrete columns for example, may in some cases be reduced to three or four highly summarised line items in the bill. This is an extraordinary waste of valuable information as well as being a source of on-going dispute amongst the members of the greater project team.

There are other areas where conventional planning and cost control techniques fail, for the same general reason. Thus, in the absence of a well-specified, shared baseline work scope, it is impossible to establish a direct or useful correspondence between cost and planning systems. So the cost and schedule dimensions of the project get out of sync, tell different stories, and provide different feedback to management, generally adding confusion to the picture.

Also, in the absence of a common specification of the scope of the project, it is more or less impossible to carry out useful analyses of the joint cost and schedule impacts of change proposals or other issues. One could go on …

The point is that any given construction project manager, starting out on a new project, equipped with all the technologies and management tools the modern industry has to offer – in particular, working with the poor quality of information provided by drawing-based design processes – faces a less than 50/50 probability of success. The strong likelihood is that, for all the stress and anxiety that he – and all the other managers on the job – will endure, he will still fail to deliver the thing on time or within budget.

BIM changes all this.

4 The solution

4.0 Introduction

Chapter 3 considered the operational-level problems caused by poor-quality drawing-based design information, and inadequate communications in construction. BIM-based design techniques promise to do away with these problems. BIM models generate dramatically higher quality information than conventional drawing-based techniques. And BIM standards and protocols lead to significant improvement in communications between firms, by enabling this inherently systematic and well-structured information to be reused directly in different computer systems.

This chapter introduces the ideas and technologies behind BIM, in advance of a more detailed discussion in Chapter 5. The main objective of the present chapter is to provide a reasonably detailed outline of the operations-level benefits that can be expected to result from the implementation of BIM-based design techniques. The discussion follows the same structure as that used in Chapter 3, the core aspects of which were illustrated in Figure 3.1.

4.1 General features of BIM-based design

In the BIM approach to building design, the designer creates a computerised three-dimensional model of the proposed building in virtual 3D space. He or she does this by inserting 'intelligent' virtual components, at precise orientations, into precise locations in the model. As the designer builds the model, the system builds up a sophisticated internal database recording the details of each of the components used. BIM components can carry many attributes in addition to their geometry and location. These attributes can be used to simulate the physical nature and related behaviours of the materials from which components are made, including their structural, acoustic and thermal properties. They can be used to simulate the way components interact with each other in the model. The information recorded about components can also include economic characteristics such as the unit cost, manufacturer identity, and planned erection dates of individual components. The database also records the details of every action carried out on the objects it contains, so that changes and other transactions can be tracked accurately throughout the design programme.

BIM components are parametric, in the sense that each component has certain variables or parameters associated with it that control its behaviour. For example the width to height ratio of a particular component might be set parametrically, so that whenever the user alters one of the dimensions the other will change according to that ratio. Relationships between components can also be specified parametrically so that if one component changes, others with which it is associated will also change according to the relevant parametric rules. Very complicated relationships can be created between different components using these parametric attributes. This enables BIM models to be used to generate remarkably detailed, realistic models of even the most complex buildings.

Working with BIM methods, when the model is complete, the information it contains can be passed to other parties in a variety of very efficient ways. If drawings are required they can easily be generated, almost like photographs of the model. Schedules can be created simply as queries against the database of components that make up the model. And, crucially for the longer term, the details of many types of components and component assemblies can be exported directly to suppliers' computer systems, even to manufacturers' computerised cutting, boring and shaping machines.

In summary, by comparison with conventional drawing-based design, BIM models provide a number of key benefits, including:

- explicit representation of the objects being designed; no dependence on cryptic forms or symbologies;
- inherent coordination of details between different views of the same component;
- direct, unambiguous association of many different types of data with selected components, resulting in extremely data-rich models;
- easily generated 3D views, complex section views, rotations, walk-throughs and such like, to enable complex objects to be designed efficiently and understood intuitively.

The model created using the latest BIM tools is a powerful, non-cryptic, flexible and rich encapsulation of the designers' intent. It offers enormous benefits to all the key players: the client, the architect, the members of the larger design team and the contractors.

Three broad types of problem with conventional design information were identified in Section 3.2:

- the typical client's inability to visualise the design accurately;
- the difficulty of integrating and coordinating cross-disciplinary design information;
- the limited ability of constructors to visualise in detail the designer's intentions.

All of these result from the use of drawings – highly stylised, abstract, cryptic, discipline-specific forms of representation – to convey the designer's ideas and to

46 *The solution*

guide construction. These problems are greatly reduced when the design is based on the use of BIM models and supporting techniques.

4.2 The advantages of BIM-based design information

The main way in which BIM can help during the design phase is by dramatically reducing the project team's dependence on drawings to communicate design ideas. The key thing here is the ability to replace lines with components; line-based drawings are ambiguous in a way that modelled components are not.

4.2.1 Advantages of BIM in design production and administration

It would be wrong to suggest that BIM dispenses entirely with the tedium of administrative work associated with a typical design programme, but it certainly helps. For the individual practice, the key advantage of the BIM approach lies in the fact that all of the relevant information is generated by and contained in that firm's single design model or database. This means that all views of the model, all drawings and schedules, and all other outputs should be inherently consistent with each other. This in turn means that the need for detailed document-by-document checking should be greatly reduced.

Patrick MacLeamy, CEO of HOK Architects, uses the diagram shown in Figure 4.1, primarily to illustrate how BIM working benefits the client. In the traditional approach, the bulk of the effort – in man-hour terms – is expended

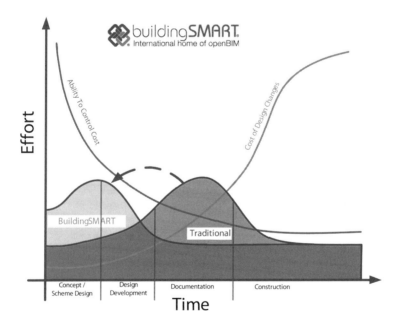

Figure 4.1 MacLeamy curve

during the draughting/documentation phase of the project. At this stage of most projects the design team's ability to control the cost of the building has more or less bottomed out; the key decisions have been made and the cost of any design changes from here on rises rapidly. MacLeamy suggests that it would be significantly beneficial to the client if the peak of design effort could be moved back behind the intersection of the 'cost of changes' and 'ability to control costs' curves. It certainly seems plausible that by investing more effort earlier, into these true design phases of the project, a better thought-out building should result. This provides more time for designers to do what they most enjoy doing – solving design problems. And crucially, it also helps to move the high-value man-hours to the front of the project.

The greatest downstream benefit of the BIM approach in design lies in the possibility of consultants being able to deliver fully coordinated, dimensioned, detailed designs, as the basis for procurement of main contracts and, particularly, specialist trade contractors. The currently prevalent practice of awarding structural and M&E contracts in particular, on the basis of scheme design or less, is deeply unsound. It is reasonable to expect specialist contractors, as they have always done, to produce shop drawings and method statements. But to expect those firms to produce professional quality, coordinated, detailed designs is unrealistic and potentially contrary to the client's interests. BIM techniques, including online access to complete vendor data from equipment manufacturers, make it possible for the consultants to carry out these services properly, professionally and, for them, profitably. (This topic re-emerges in Chapter 9.)

4.2.2 Advantages of BIM in communications with the client

In the BIM approach, the design can be presented to lay viewers, such as client organisations, as a photo-realistic, walk-through – 'what you see is what you get' – model. This explicit, non-cryptic method of representation greatly improves the client's confidence in his understanding of the scheme and enables early decisions to be made with much greater certainty than is usually the case. Subsequent stages of the project can then proceed smoothly, with a minimum of client-instigated design changes.

When a client views a design, he wants to see as clearly as possible how the designer is proposing to solve his problems. He wants to see the solution in his terms, or at least in terms that are clear to him. He wants to see the whole solution; not just how the building will look, but also how much it will cost and how long it will take to construct, and so on. And he wants to be presented with the solution in such a way that he can interrogate it and interact with it.

Component-based BIM models allow exactly this sort of dialogue to take place. The client can be presented with images that are, in a sense, better even than photographs of the proposed building would be. He can immerse himself in the model, walk around it, see views from outside looking in, from inside looking out. He can see it as it would look at different times of day, at different times of year. He can see simulations of people moving through the building. He can swap features like surface finishes at the click of a button. He can drop a virtual key on a virtual marble floor and hear the sound reverberate around his virtual foyer.

That deals with how the building will look and feel, even to a lay viewer. However, one of the crucial features of BIM models is that they can include a wide range of information types that might be regarded as being supplementary to the basic design. So in addition to the building's geometry and such like, a BIM model can also show the user how much it would cost and how long it would take to construct the building, depending on which particular architectural solution he chooses. And this can be done simultaneously, as he explores the architectural options.

Perhaps the most important advantage of the BIM approach to the client is the dramatic improvement in certainty of outcomes – cost, schedule and quality – that he can expect. The model acts as a baseline for the project. It enables quality standards and cost and programme targets to be established clearly and accurately. As work proceeds, any variation from those objectives can be picked up and responded to quickly and precisely. So changes should be rare and their impacts should be kept well within agreed tolerances.

4.2.3 Advantages of BIM in communications with the design team

In addition to the improved clarity of representation of the design intent, the individual drawings generated by BIM models are inherently internally consistent. There can be no mismatches or internal contradictions in the content of any individual document or related sets of documents generated from a given model. This dramatic increase in clarity and consistency leads to a much more efficient design production process within the individual firm's design team.

Compared with drawing-based design, it is also relatively easy to coordinate the design contributions of different disciplines by incorporating them and testing them in a shared BIM reference model. Eliminating visual and architectural ambiguity is a large part of the appeal of BIM as a means of communication with the client. This capability is also obviously important in communications between different members of the design team. However, the power of a BIM model to act as a means of accurately exchanging precisely specified, structured data, between the design disciplines is its most important strength in this context.

Amongst other things, this means that many fewer multidisciplinary design review and integration cycles are required to complete the design of the building, again leading to a much more efficient overall design process. Any form of multidisciplinary design is an inherently iterative process. The lead designer thinks of an idea and draws it up; a support designer, a structural engineer, for example, considers this idea and thinks of changes, which he draws up and feeds back to the lead designer. The lead designer considers the proposed changes, thinks of changes to them, which he draws up and feeds back to the support designer. The support designer considers these proposed changes to his proposed changes ... and so on.

In each of these exchanges, two possible types of error occur. First, the designers may not interpret each other's lines or symbols correctly: a line is construed as representing a pipe rather than a cable, for example; and secondly, they may transcribe each other's images inaccurately: line of length 2.003 m rather than 2.000 m, grid orientation 183.00°, rather than 183.06°. These sorts of issues are compounded enormously as the number of designers grows and as the complexity of the building increases.

BIM models overcome these problems in two ways. First, by presenting the various options in a complete, explicit form, they reduce the number of iterations needed to arrive at the optimum solution to any given problem. And secondly, by providing a single data exchange environment, they reduce the potential for error in each iteration. The result is a streamlined, highly profitable design process, made even more so by the virtual elimination of low-value drafting activities.

4.2.4 Advantages of BIM in communications with the contractors

The third audience for whom effective design communication is important is the construction team. The contractors need high-quality, unambiguous information to be delivered in a smooth, programmed stream. They use the information for two main purposes: to procure specialist trade contracts, and to plan and manage the construction operations of those contractors. The information generated by a BIM-based design process is, almost by definition, clear, unambiguous and complete, for whatever purpose it is to be used, so it readily satisfies the principal requirements of the contractors.

In BIM working, the flow of information between the design team and the contractors can relatively easily be organised so as to be timely and smooth flowing. There should be no need for documentation to be delivered late or sporadically in big, indigestible lumps. As noted above, BIM working enables (and requires) more of the detailed design decisions to be made earlier in the overall design process. This, together with the improvement in client decision making and closer communication amongst the members of the design team discussed above, should make it possible to plan and manage the release of design information far more effectively than is traditionally the case. Late, lumpy packages of information should be a thing of the past.

To reinforce the point made earlier, it is highly desirable in almost all forms of construction for the maximum possible proportion of the work to be awarded on a competitive, fixed price, lump sum basis. By eliminating the 'blind man and the elephant' problem, BIM models support this approach to contract procurement far more effectively than conventional design documentation. They also enable the design team to achieve whatever level of dialogue they require with construction managers and specialist contractors and suppliers, with minimal prejudice to competitive objectives of the client's procurement strategy. BIM enables the optimal combination of competition and collaborative working to be achieved.

4.3 The advantages of BIM in contract procurement

In a BIM model there exists a digital equivalent for every one of the components that make up the physical building. The model takes the form of a powerful database which can hold a great deal of data about each of these components. The data held includes information about the component's classification, as well as the details of its physical properties, its geometry and its location in the building. The database can be queried in a wide variety of ways. So, for example, it would be quite simple to extract a complete, detailed schedule of components, grouped by classification or by trade and by location. A copy of this schedule, in database or spreadsheet format, together with the relevant drawings – or even better, the relevant model or model section – can then be sent to the bidding contractors, for them to price and return.

The key benefit of the component schedule is that it provides a definitive, verifiable statement of the scope of work of any given trade contract, in terms of components to be installed into the building. Each of these components can be located, identified and itemised by all of the bidders, so their tenders will be exactly comparable to each other, at the component level of detail. There is no scope for gaming the process, and given that the design stage, as described above, delivers a complete design, there is no opportunity deliberately to bid low in pursuit of claims. Contractors compete on the basis of their ability to perform the work most efficiently, that is at lowest cost, rather than on their ability to chase claims.

In this scenario, the two key objectives of a competitive market are achieved: lowest price and best performer. And because the contractors don't have to worry about predatory bidders, they can pitch a reasonable price, including a reasonable provision for overheads and profit.

The strategic, industry-level importance of this capability is almost impossible to over-emphasise. Like most other sectors of the economy, construction is a highly competitive industry. However, competition in construction is focused almost entirely on winning projects, not on delivering them. Conventional documentation which contains so much information that is incorrect, unclear, inconsistent, uncoordinated and incomplete, forces contractors to take extraordinary measures to win work. BIM documentation, on the other hand, both requires and enables contractors to bid competitively for the actual construction work. In such an effectively competitive market, the price of buildings will fall, but the cost of building will fall faster, as economic theory demands.

4.4 The advantages of BIM in construction management

Ideally the BIM model should be created during the design phase of the project, as a deliverable from the design programme, and used as outlined above, to support design communication. However, even if a BIM model is not produced during design, it is a relatively easy and very cost-effective exercise to develop such a model for procurement and construction purposes. Once created, a BIM model provides a unified, coherent representation of the building that everyone involved in the project, and everyone with an on-going interest in the operation of the building, can use and benefit from.

BIM models will assist in the management of construction operations in two main ways. The first of these involves simply using the graphical power of sophisticated 3D models as a means of visualising how the building fits together. This includes clash detection and construction simulation exercises, as well as cutting special views of parts of the building to assist in the solution of particular construction problems. Strictly speaking, this is not using the information component of BIM models, but it's likely to be the application area that drives the use of BIM onto the construction site.

The second way in which BIM will assist in construction is in the dramatic improvement in the sheer quality of the design information that's created in a well-made BIM model. This effect will be much more profound and more important in the long run, in two respects. First, BIM models will provide information that will be completely trustworthy, in the sense referred to earlier. This information will be correct and complete. It will enable project teams, without any need to check, to use the information they are provided with, directly and with complete confidence; no need for personal assessments or subjective judgement, just get on and use it.

This leads to the second feature of BIM information – call it data now, because that's what it is: clean, well-specified, computable data. Computability is crucial because it removes the need for human intervention in the flow of the idea from its point of origin in the architect's mind to the point of its application in the hands of the artisan. More prosaically, it means that the precise data set associated with any given component can be passed automatically from the point at which the virtual component is inserted into the BIM model, through all the systems controlling detailed specification, material take off, procurement, manufacture, assembly, storage, shipment to site, handling on site, installation and testing and handover. The entire end-to-end history of the component is captured and managed seamlessly in the BIM model and related applications – because the information takes the form of properly specified, computable data. This is not a simple evolutionary development; it's a point of abrupt, discontinuous change.

It may be difficult to grasp the significance of this, to see quite what it means to suggest that the actual nature of information can change in this sense, and how that really impacts on the way firms can use it. But something very similar to BIM happened in aircraft, car and consumer product manufacturing over the past 30 years, with the introduction of CAD/CAM and computer integrated manufacturing (CIM) and supply chain management systems. A similar transformation ensued in the retail sector when electronic point of sale (EPOS) systems were introduced about 30 years ago. It's taken more or less a whole educational generation, but today's manufacturing and retail sectors are fundamentally different to what they were then. These were both low-quality, low-profit, high-cost industries, driven by seat-of-the-pants judgements of individual engineers and managers. Today's manufacturing and retail industries are high-quality, high-profit, low-ost businesses driven by high-quality, properly specified, computable data.

BIM is an information technology that is specific to construction, just as CIM is to manufacturing and EPOS is to retail. However, there is a great deal that construction can learn from the experience of these other industries, in the way in which their particular technologies were introduced and their effects on industry performance and structure. These issues are explored more fully in Chapter 8. First, though what are the specific ways in which BIM will improve the predictability and profitability of the project construction phase?

4.4.1 Project management issues

It was pointed out in Section 3.1 that projects over-run their cost and schedule targets for two broad reasons: targets set incorrectly in the first place, and inaccurate status assessments in the course of the work. These two things happen largely because of the poor quality of information available on a conventional project, on which to base the scope of work of an individual contract package, or of the project as a whole. The information is inadequate and untrustworthy in all the ways described in Chapter 3.

To overcome these deficiencies, the project planners, cost engineers and others must apply their personal, subjective, experience-based judgement to guess the correct meaning and to fill in gaps in the available material. This in turn gives rise to the problems outlined in Sections 3.4.1 to 3.4.4: excessive subjectivity, inappropriate use of planning systems for forecasting, lack of connection between plans and reality at the production level, and lack of responsiveness in project control tools generally.

The picture is transformed when the information on which the scope of work is based is derived from a BIM system. In this scenario, no guesswork is required, quantities are known precisely, so targets, based on accurate detailed historical rates of production, can be set accurately and with confidence. Equally, progress can be assessed by accurate measurement of the numbers of components installed in a given period. Forecasting cost and schedule outcomes is then a matter of simple arithmetic. And deviations from the plan, should they occur, can be spotted early and responded to while they can be rectified at minimal cost and with least disruption.

The quantity of data generated in managing the project at this level of detail is prodigious, of course. But, to repeat, the data is highly structured and systematic. A particular beam or window in the BIM model, reflected in the BIM-based production programme, can easily and unambiguously be recorded as being complete when it is seen to have been installed on site. It's a binary, black or white, done or not-done assessment, so no dispute or uncertainty exists.

For the moment, approaches to BIM-based project management tend to embody conventional cost planning, estimating, quantity surveying, planning and other project control techniques. Increasingly however, these are likely to be superseded by methods and systems that more effectively assist the entire construction team to optimise levels of output and particularly the labour productivity of all the firms involved in the work. Section 4.5 provides an outline description of such a system.

4.4.2 Construction as an assembly process

People tend to think of construction as being an industrial dinosaur, conservative and slow-moving, reluctant to innovate and viscerally resistant to change. That picture is actually very far from the truth. Over the past 50 years the industry has innovated continuously and has changed dramatically, both in structure and in its modes of operation.

In the period since the Second World War a number of particularly important, closely related trends have been at work:

- There has been a huge increase in the numbers and variety of standardised construction products, materials and components.
- There has been a significant reduction in the use of craft-based working on site. This has been replaced by relatively low-skilled, but highly specialised, assembly and installation site processes. These services are increasingly provided on a labour-only, sub-contract basis.
- The main contractor constructs very little; specialist sub-contractors carry out most of the actual building work.
- Concern for the health and safety and other aspects of the wellbeing of the workforce has increased significantly.
- The architectural profession is becoming increasingly focused on pure design. The traditional role, including detailed coordination of the other disciplines' designs, and application of in-depth knowledge of construction techniques and materials is receding in importance. Design for craft working is disappearing; design for manufacture and assembly is beginning to predominate.

Extrapolating these trends, and to some extent looking at the experience of other industries, it would seem reasonable to suggest that the construction site of the future will increasingly take the form of a highly controlled, industrial assembly site. There will be no wet trades. Nothing will actually be made on site – not even concrete. Buildings will be constructed entirely from standard components and pre-assembled modules, fabricated off-site, in factory conditions. The site itself will also look more and more like a factory.

The management emphasis in this scenario will be on maximising production rates and on optimising the productivity of individual work crews; ultimately, maximising output per man-hour of labour time. The role of the main contractor on site will continue to be to procure, facilitate and coordinate the activities of large numbers of specialist sub-contractors. However, his priority, much more explicitly than today, will be to enable the individual specialists to perform as efficiently and as productively as possible. The key measure of project success will be the profitability of the specialist sub-contractors; if they make money, everyone makes money. So the focus of all concerned will be on managing production on the job site.

4.4.3 Building owner/occupier

As the contractors install individual components and items of equipment in the building, the details of each of these events, facts like the date, time, work crew, testing and commissioning information and so on, are all recorded in the BIM model, simply as part of the activity of managing the production process. The result is an 'as built' version of the BIM model, generated automatically, continuously during the course of the work. This is a huge improvement on the scrambled, shapeless bundles of paper 'as builts', operation and maintenance manuals and health and safety files that are cobbled together at the end of most conventional projects.

The BIM 'as built' model allows the owner/occupier to simulate, test and generally optimise the functionality and performance of the building throughout its life-time. It becomes a powerful asset-management tool, which enables the owner to truly maximise the return on his investment in the building.

4.5 Production management in construction

In the production management approach, the key thing to know is how much physical production is planned for a particular period and how much is actually being achieved on a detailed, on-going basis. This can only be done by measuring production levels accurately; measuring the numbers of individual components installed, component-by-component, day-by-day, throughout the project. This, in turn, can only be done if each component can be uniquely identified and tracked through its entire journey from design model, through procurement and manufacture, to site and into its ultimate position in the finished building. This is exactly what BIM, uniquely, makes possible.

Conventional, CPM-based project management deals mainly with the activities that have to be carried out in order to complete a given project. Production management deals with the physical output generated by those activities. Thus, a traditional CPM-based activity plan shows the work that will be under way, day-by-day or week-by-week during the project. A production plan shows the level of physical output that must be generated day-by-day, or week-by-week throughout the project in order to complete the job on time. The distinction between these two approaches may seem subtle, but it's actually crucial.

The production management approach overcomes all of the problems associated with project management discussed in Section 3.4:

- It eliminates dependence on intuitive, subjective definition of work scope and progress assessment.
- It connects management directly to events at the work face.
- It enables dramatically improved forecasting and trend monitoring.
- It provides a comprehensive framework within which to gather, analyse and re-deploy performance data from one project to the next.

The solution 55

The production management approach to the management of construction is illustrated in the diagram in Figure 4.2. In broad terms, it works as follows.

- The starting point for the production plan is a detailed schedule of the components of the building in question, taken from the Building Information Model, or if a model is not available, taken off from drawings, in the form of a 'smart' bill of quantities. Each component is identified according to a catalogue of unique component codes – similar in concept to the retailers' UPCs – and its location and quantity are recorded.
- Each component or group of components is then allocated to one or more construction packages. The man-hours and/or monetary cost of installing each component are assigned according to the package schedule of rates. In this way the 'planned value' of each installed component, thus of the package as a whole, is established – before construction gets under way.
- The key point about this step is that it establishes the 'value' of all the various components of the building in common units: man-hours or monetary cost. This is the basis on which all subsequent calculations are carried out. Reducing everything to a common basis of value enables comparison and aggregation of completely different types of operation, brickwork, carpentry or plumbing for example, over time on any given project, and importantly, across different projects.
- The planned progress curve for the package in question is generated by spreading the man-hours or costs across the package, or sub-package, duration according to a simple production function. This curve shows the level of output that must be achieved every day or every week in order to achieve the agreed package end date. (The assumption here is that the specialist contractor mobilises an optimum crew to site, aims to get them up to their optimum rate of output as quickly as possible and to keep them at this rate for as long as possible, until the work is complete.)
- When work gets under way the actual installation of individual components is recorded, as it happens, preferably on a continuous, daily basis. Every time the contractor completes the installation of a component he 'earns' the man-hour or monetary value of that component.
- This information is added to the production plan to generate the 'earned value' of the work performed. The actual percent complete of the package can then be calculated as the ratio of the earned value to the initial total planned man-hour or monetary value of the package. This is plotted as the actual progress curve. And obviously, as all achievement is denominated in a single unit – man-hours or money value – packages can be grouped and totalled to give group and overall project values and curves.
- It is highly desirable that when the fact of a component installation is recorded, the man-hours used in its installation are also noted. This enables labour usage, productivity and effectiveness to be monitored and managed.

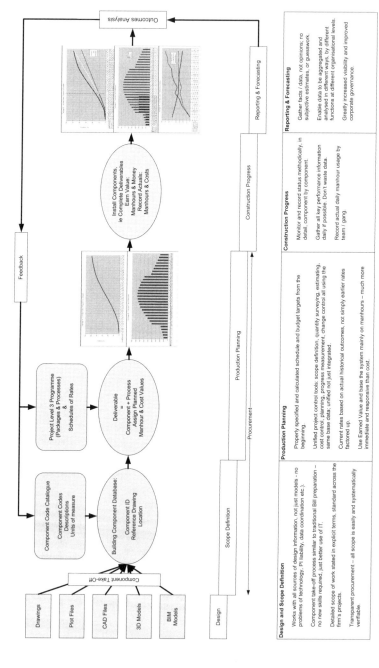

Figure 4.2 The production management approach

In general the information required to drive production management is neither new nor particularly different. 'Dumb' bills of quantities have been in use in construction for over a century. However, as paper documents they lack detail, flexibility and authority. A 'smart' bill, particularly one generated from a BIM model, retains all of the information about the building at the most detailed component item level. Unlike the dumb traditional bill, the smart bill is flexible in that it can be used for many purposes other than just tendering and valuation and it is authoritative in that every item it contains can be challenged explicitly, and can thus be verified definitively.

The progress information used in production management is a simple extract from the trade supervisor's daily diary. The man-hour usage information can be obtained from the same source or from the contractor's time recording system. And of course, the system can be used to generate turnaround documents for the capture of this information or to drive hand-held data capture devices or similar.

This is just one example of the sort of construction production managment system that might be implemented as an overlay to a BIM platform.

4.6 Conclusion

BIM enables design teams to create information that is of dramatically higher quality than that produced using conventional design techniques. This quality improvement will have its effect in two ways. First, simply because they can trust the information they are working with to be accurate and reliable, project teams – clients, designers, contractors and suppliers – will all experience huge benefits in their internal information activities and in the way they communicate with each other.

A BIM model can overcome most of the most serious failings of conventional drawing-based design. It provides greater client certainty earlier; improved consistency and easier coordination of design documentation; improved, complete procurement documentation; much more powerful construction and project management tools; and much more valuable 'as built' and record information for owner. The result will be substantially more profitable firms of all types in the sector, delivering projects a great deal more reliably.

The second, more profound set of effects will come when firms in the industry start to take advantage of the fact that the information generated by BIM systems is fully computable. This will enable the huge variety of building-related data to be passed directly, without any need for checking or re-keying, from system to system, in a single continuous stream, all along the construction supply chain. Chapter 8 will describe how, over the past 20 years or so, particular industry-specific information technologies have transformed the manufacturing and retail industries, by enabling their members to achieve exactly this – fully digital, end-to-end data. It won't take quite so long, but we can confidently expect BIM to have a similarly transformative effect on construction.

5 The origins of BIM in computer-aided design

5.0 Introduction

Today's Building Information Modelling systems – the software tools used to create BIM models – all originated in mainstream computer-aided design. So, before moving on to a description of the technical capabilities of BIM systems, it will be useful to discuss the key features of computer-aided design in general and to outline briefly its short, interesting and surprisingly complicated history.

5.1 Terms clarified

To begin with, it will be useful to clarify the meaning or usage in this book of some of the key terms used in the domain of computer-aided design. The expression computer-aided design (CAD) refers to the whole spectrum of ways in which computers can be used to assist in the design, drafting and engineering work associated with the production of a new product or refurbishment of an existing object.

5.1.1 Drafting versus modelling systems

A distinction is drawn between two broad types of computer-aided design tools: drafting systems and modelling systems. As the name suggests drafting systems are used to create, edit and print/plot two-dimensional drawings. Their purpose is primarily to make the drafting activity more efficient. Such a system contains nothing apart from the lines and curves entered by the operator. Early computer-aided design systems were almost entirely of this type. As Weisberg points out:

> These systems were marketed predominately on the basis that they could reduce current operating costs. If you had a drafting department with 20 drafters, buy one of these systems, run it around the clock and you could get the same amount of work done with perhaps 10 or 12 people. In some cases, productivity improvements were truly spectacular, especially within organizations that did a lot of repetitive work.[1]

---

1 Weisberg, D.E., *The Engineering Design Revolution: The People, Companies and Computer Systems that Changed Forever the Practice of Engineering.* pp.2-9, 2-10. http://www.

Modelling is quite a different type of activity. In this approach, the operator creates a computerised three-dimensional model of the object being designed. This can be done using drawing techniques, for example by drawing out two-dimensional objects and sweeping or extruding them to create three-dimensional forms. Alternatively, and more effectively, a model can be created using true modelling systems in which three-dimensional objects or components are inserted into computerised three-dimensional space. It is also possible to create a model using a combination of the two techniques.

Once it's complete, such a model can be used to generate any required view of the object being designed, including arbitrary sections and perspective images. The key point about a model is that being a true, complete likeness of the object, it contains all of the geometry necessary to generate these views. Generated drawings and model views are rather like photographs of the object.

5.2 CAD application areas – key challenges

This chapter is concerned with the use of computers in the design of process facilities, of engineered mechanical products and of buildings and civil engineering works. There are many other important areas where computers are used in the design of products, including the design of electronic components: printed circuit boards, microprocessors and suchlike; and in fashion manufacturing, including fabric and shoe design. These application areas are outside the scope of this book.

5.2.1 Process design

This area includes facilities like production plants and ships which comprise large numbers of components. Their geometry is relatively simple, but their engineering is highly complex. In process engineering, much of the design effort is carried out through the medium of schematic or logic diagrams, such as process flow diagrams, piping and instrumentation diagrams, electrical single line diagrams and so on. The challenge here is to ensure that the representations of components in all of these documents, as well as the geometry – general arrangement drawings and such like – remain synchronised and consistent with each other as the design is developed.

5.2.2 Mechanical design

This includes the design of objects like aircraft, cars, boats and consumer products. Typically, these objects are made up of relatively small numbers of individual components. Their geometry and engineering are both generally complex. The key challenge with mechanical design systems is ensuring that components and assemblies of components are fully specified and defined, so that the data

---

cadhistory.net/ (retrieved: 24 October 2010). David Weisberg's encyclopaedic insider's knowledge of the history of CAD, particularly mechanical CAD in the USA, is fascinating reading.

representing them can be relied on to drive the machine tools and assembly systems used in their manufacture.

> For digital design models, this means that the geometric model intended to govern a digitally based production process must be unambiguously defined (one that doesn't have floating one-sided planes or unattached lines, for example), and it must successfully characterise the anticipated physical object.[2]

It is almost intuitively obvious that, whereas an experienced human craftsman can accommodate approximately accurate instructions and can, intelligently, bridge gaps in those instructions, the same is not true of numerically controlled (NC) machines. Any NC system, such as those used to create surface forms or to trace the paths of cutting tools, must be provided with precisely accurate, complete information about the object being manufactured.

5.2.3 Architecture, engineering and construction (AEC) design

Buildings comprise large numbers of components, but their geometry and engineering are relatively simple. One of the main systems challenges in attempting to use computers in support of AEC design is dealing with the concept of design intent, an approach to design that is inherent in much of contemporary AEC practice. In this approach, rather than providing explicit, detailed instruction as to how a particular feature should be constructed, the designer describes the effect he is seeking to achieve, and leaves the actual implementation decisions to the contractors and tradespeople who will actually carry out the work.

Drawings produced for this purpose are often only indicative and are, almost by definition, incomplete and undetailed. Even if it has been created using computer systems, it is generally the case that this sort of information is issued only as paper drawings, or in the form of uneditable drawing files – so that the user is actually forced to interpret the originator's design intent, and essentially, to create his own version of the design idea. This is significantly more onerous than the traditional requirement for specialist contractors to provide shop drawings and method statements; or means and methods, in North American parlance.

The contrast between mechanical and architectural design thinking is dramatic and fundamental. As Jon Pittman, VP, Building Construction and Management Solutions Division, with Autodesk Inc. put it:

> To mechanical engineers, 'design intent' referred to a set of very precise dimensions, constraints and parameters that drove the design concept. Their focus was on ensuring that the manufacturers would fabricate the products according to absolutely defined tolerances and specifications – with no

---

2   Schodek, D., Bechthold, M., Griggs, K., Kao, K.M. and Steinberg, M., *Digital Design and Manufacturing: CAD/CAM Applications in Architecture and Design*. Hoboken, NJ: Wiley, 2005, pp. 314–5.

ambiguities about what was manufactured... Architects sought to express their design intent more broadly – clear enough for a contractor to construct the building without explicitly providing instructions for how to do so. Why the purposeful ambiguity? According to some, ambiguity is necessary in order to minimise the architect's own liability in case something goes wrong during the construction process. Others say that purposeful ambiguity allows our industry to tap into the distributed intelligence of the community i.e. that the collective knowledge of how things get built as embodied in designer, builder, manufacturer and tradespeople, is far richer than the knowledge embodied in any one individual or group. Further, limitations on the architect's compensation made finding efficient ways to depict the building necessary.[3]

5.2.4 CAD – a spectrum of applications

For the purpose of this book, the overall concept of computer-aided design, referred to here as CAD, will be considered as encompassing a spectrum of applications. This classification is based primarily on the format of the information generated in the design process and its intended or potential subsequent usage.

- Computer-aided drafting – referred to here as CAD. systems may support 2D and 3D drafting, not usually true modelling. This is the most basic form of computer-aided design, in which the main outputs are dumb drawings, which can take digital or paper form. These are read-only documents, which are not intended to be edited or otherwise re-worked by other users. All subsequent usage of this information requires a significant degree of human judgement, interpretation and evaluation. No substantial interchange of reusable digital data takes place.
- Computer-aided drafting and design – CADD – in which drawing, modelling and other design activities are closely integrated. CADD data, in digital form, is exchanged between drafting and other applications, such as engineering analysis and material scheduling systems. The deliverables can include 2D drawing files, 3D model files, engineering calculations, bills of materials and other outputs. All information is created with the explicit intention that it can be reused by its recipients in other computer systems. The information is therefore 'trustworthy'; accurate, complete and conformant with agreed standards. Such information does not need to be checked for meaning, accuracy or completeness by its recipients, but decisions on its use involves a degree of human intervention.
- Computer-aided design/computer-aided manufacturing (CAD/CAM), in which fully engineered drawing and engineering CADD data is passed directly to manufacturing, in digital form, for use in computerised numerically controlled (CNC) machines, or other types of automated manufacturing

---

3 Pittman, J., 'Building Information Modelling: Current Challenges and Future Directions', in B. Kolarevic (ed.) *Architecture in the Digital Age – Design and Manufacturing*. Abingdon: Spon Press, 2003, p.255.

facilities. In addition to the CNC data, outputs may also include paper and other graphical forms, including computer-generated prototypes. All information is created with the explicit intention that it can be used to guide the operation of numerically controlled machines, without any significant human intervention.
- Computer-integrated manufacturing (CIM), in which CAD/CAM data is used as input to the design and operation of manufacturing processes and systems.

5.3 A brief history of computer-aided design

As with many other instances of major innovation, the history of CAD interweaves ground-breaking intellectual breakthroughs with substantial technological advances and astute commercial opportunism. Some people contributed remarkable fundamental mathematical discoveries; others developed wonderful computing machines with which to explore this new knowledge, and a third group saw and grasped the opportunities to combine these two to create an industry that had never previously existed. All three groups are crucial players in the 50-year history of computer-aided design.

A second pattern worth noting is the extent to which the fundamental insights that enabled these innovations are often associated with individual key researchers. This may be a matter of history being written by winners, but the story of these developments demonstrates that, although they were usually working in teams of highly gifted people, the creative spark almost always seems to originate in the working of a single mind.

Three identifiably separate groups of organisations were involved in the development of computer-aided design systems. The fundamental research was done mainly by government-funded academics in universities in the USA and the UK. Most of the early applied work was done by large industrial users of CAD. The more recent, commercial exploitation of CAD was all carried on by successive generations of entrepreneurial CAD software vendors.

5.3.1 Computer-aided manufacturing

It may seem a digression, but before embarking on the CAD story proper, we need to consider briefly the story of computer-aided manufacturing (CAM), a complementary technology that came slightly earlier than CAD. Machine tools like lathes, drill presses, routers, milling machines, dies, planes and such like have been used for centuries in the wood, stone and metal-working industries. Traditionally these have been controlled by highly skilled operators using the chucks, wheels and levers of the machines to manoeuvre the work-piece and to aim and hold the tools in place as they worked. Various methods were tried of automating the control of these machines, to ensure that they produced exactly the same precisely finished, interchangeable product every time. Methods including devices like templates, cams and even punched paper tape, similar to the Jacquard loom, had been tried, but with only mixed success.

During the Second World War, methods of controlling machines using elementary programs were developed. The programs – the particular sequences of tool positions, movements, speeds and so on needed to machine a particular object – were keyed by hand on to punch tape or cards. Then, instead of a human operator, the numerical program was used to drive servomechanisms which in turn monitored and controlled the positions of the work-piece and the tool through the manufacturing process. This early form of numerical control (NC) was the first effective example of the craftsman's production intelligence being programmed into a machine. Henceforth, the human operator's only role was to initialise the machine at the beginning of each production batch and to clear it down at the end.

The original NC concept took off only slowly, because it was so difficult and took so long to generate the machine control programs. The breakthrough happened when computers were introduced into the process and simple NC machines became CNC – computer numerically controlled – machines. This was largely due to the efforts of two of CAD's great innovators: Doug Ross[4] who developed the Automatically Programmed Tool (APT) NC programming language, while at MIT in the 1950s; and Patrick Hanratty, founder of Manufacturing and Consulting Services (MCS), creator of ADAM, released in 1971, the first system that integrated design, drafting and CNC manufacturing – the first commercially available CAD/CAM software.

The crucial feature of CAD/CAM working is that the design process works at the same level of detail as the manufacturing process. So, details such as chamfers, countersinks, bolt and bolthole threads must be present in both. Specifically the design of the object to be machined must be accurate, precise to tolerances of thousandths of an inch, and absolutely complete. Machines can't guess, or estimate the 'design intent' in the way a skilled craftsman can.

However, '... "numerical control is one of the most important basic innovations of our century ... it has gone far beyond the original cutting-machine tools and has revolutionised manufacturing and other areas of human activity." It has totally changed how engineering design is practiced and has been a major element in the increase in industrial productivity we have seen during the past decade.'[5]

5.3.2 Drawing with a computer: the problem described

The first challenge with CAD is understanding how objects like buildings or cars or ships are to be represented in computers. Things like invoices, delivery

---

4 Ross, D.T., 'Origins of the APT language for automatically programmed tools', ACM SIGPLAN Notices 13 (8): 61–99, 1978.
5 Kochan, D. (ed.), CAM: Developments in Computer-Integrated Manufacturing, New York: Springer-Verlag, 1986, p. 6, quoted in: Cortada, J.W., The Digital Hand: How Computers Changed the Work of American Manufacturing, Transportation and Retail Industries, New York: Oxford University Press, 2004, p. 111.

notes, payment certificates and such like are relatively straightforward. They are essentially just numeric data – uncomplicated, recognisable entities that can be input unambiguously and can be computed – sorted, added and subtracted – in familiar software programs, like spreadsheets and databases.

The representation in computers of physical things, or as the design process requires, internal mental images of physical things, is fundamentally different and far more difficult – largely because the way in which we see or imagine objects is far more complicated than the way in which we see or think about data and numbers. There are many explanations, but basically the human brain is thought to perceive or recognise objects in a scene in a sequence of steps: identify the edges or boundaries between objects; identify the spaces within and between them; identify the geometric relationships between the objects; identify the shadow or shades on their surfaces; and ultimately identify each of the whole, overall objects in the scene.[6]

Technical drawing involves a similar process, based on the perception or definition of the edges of objects to be drawn. So, to start a drawing, the designer selects a view of the object to be drawn and draws lines to represent its edges, as he perceives them. Next, he rotates the thing (in his head) and draws a second set of lines to represent the new set of edges perceived in this second view plane. He then performs a second rotation and draws a third set of lines to create a third planar view. The rotations are usually orthogonal to each other and, with simple items, the result is usually sufficient to reasonably fully describe the drawn objects. The graphical entities used in this process take the form of points, lines, curves and polygons, sometimes with shading applied to distinguish individual surfaces.

To enter information of this sort into a CAD system, the user first chooses or specifies a coordinate system and a plane in that system in which to work. He then needs to tell the CAD system what each line entity is, its point of origin, its length and its orientation within the coordinate system. From the earliest recorded CAD tool – a program called Sketchpad, created by Ivan Sutherland as a postgraduate student at MIT in 1962/63[7] – this information has been entered into CAD systems interactively, using computers with graphics screens and data entry devices like light pens, pucks, graphics tablets, and mice. The following quotation is from the citation for Dr Sutherland when he won the Franklin Institute Certificate of Merit:

> At a time when cathode ray tube monitors were themselves a novelty, Dr. Ivan Sutherland's 1963 software-hardware combination, Sketchpad, enabled users to draw points, line segments and circular arcs on a cathode ray tube with a light pen. In addition Sketchpad users could assign constraints to whatever they drew and specify relationships among the segments and arcs. The diameter of arcs could be specified, lines could be drawn horizontally or vertically, and figures could be built up from combinations of elements and shapes. Figures could be copied, moved, rotated, or resized and their

---

6 Edwards, B., *Drawing on the Right Side of The Brain*. London: HarperCollins, 2006, p. 96.
7 Rooney, J. and Steadman, P., *Principles of Computer-Aided Design*, London: Open University / Pitman, 1987, pp. 1–2. Still a classic text.

constraints were preserved. Sketchpad also included the first window-drawing program and clipping algorithm which made possible the capability of zooming in on objects while preventing the display of parts of the object whose coordinates fall outside the window.

The development of the Graphical User Interface, which is ubiquitous today, has revolutionized the world of computing, bringing to large numbers of discretionary uses the power and utility of the desk top computer. Several of the ideas first demonstrated in Sketchpad are now part of the computing environments used by millions in scientific research, in business applications, and for recreation. These ideas include:

- the concept of the internal hierarchic structure of a computer-represented picture and the definition of that picture in terms of sub-pictures;
- the concept of a master picture and of picture instances which are transformed versions of the master;
- the concept of the constraint as a method of specifying details of the geometry of the picture;
- the ability to display and manipulate iconic representations of constraints;
- the ability to copy as well as instance both pictures and constraints;
- some elegant techniques for picture construction using a light pen;
- the separation of the coordinate system in which a picture is defined from that on which it is displayed; and
- recursive operations such as 'move' and 'delete' applied to hierarchically defined pictures.

The implications of some of these innovations (e.g., constraints) are still being explored by Computer Science researchers today.[8]

As the citation suggests: 'Most of the elements of modern CAD systems were thus present in embryonic form in this pioneering work of the late 1950s and early 1960s: two-dimensional computer drafting; three-dimensional computer modelling; automatic analysis of the performance of designs; and at least the potential for integrating design with manufacture in CAD/CAM.'[9]

Among the advantages of CAD systems, even at this time, the most important were the reduction in errors they offered over manual drafting, the economics of reusable, stored drawing elements and the huge increase in office productivity and savings in labour costs to which they led.

So, these early, drawing-based CAD systems were used primarily as labour-saving, computer-aided drafting systems. They have been replaced with much more powerful tools in a number of industries, but it's important to note that 'Descendants of these kinds of systems, which fundamentally focus on two-

---

8   http://design.osu.edu/carlson/history/lesson3.html (retrieved 28 May 2010). The Franklin Institute Awards are among the oldest and most prestigious comprehensive science awards in the world. Among science's highest honours, the Franklin Institute Awards identify individuals whose great innovation has benefited humanity, advanced science, launched new fields of inquiry, and deepened our understanding of the universe.
9   Rooney and Steadman, p.2.

dimensional representations of three-dimensional objects, remain in common use.'[10] This is particularly true in the architecture, engineering and construction (AEC) community.

5.3.3 Early systems: wireframe, surface and solid modelling

From the initial two-dimensional drawing systems, it was a relatively simple step to the next level of CAD sophistication – wire-frame modelling. A wire-frame model is a true three-dimensional representation of a physical object in which the edges of the object in question are drawn on screen. In the system, each edge is labelled and its details recorded in a data table. The coordinates (usually in Cartesian (x, y, z) space) of the ends of each edge are also recorded in a second table. The edges are not necessarily straight lines; arcs and other well-defined curves can also be used.

The great beauty of wire-frames is their economy of data and hence the speed with which they can be recalculated and regenerated for visualisation and similar purposes. They provide a useful capability intermediate between drafting systems and true, three-dimensional CAD. Some systems enable hidden lines – lines representing edges that should not be visible in a given real-world view of the object – to be suppressed in the displayed image. (A common use of wire-frame models is in conjunction with full-blown three-dimensional modelling systems. When a large model of this type is created it may comprise so much data that it cannot easily be rotated quickly and smoothly. By reducing it down to a wire-frame form, the model can usually be rotated quickly to the desired view and only then regenerated – a much faster, smoother process.)

However, wire-frames have two major weaknesses: non-validity, in that it is possible to create a wire-frame model of an impossible object (as in M.C. Escher's drawings), and ambiguity, in that it is possible for a valid model to be interpreted as representing more than one real-world object. These flaws dramatically restrict their usefulness.

During the 1950s and 1960s development in CAD technologies was carried out mainly in universities – notably MIT and Harvard in the USA and Cambridge, England – and in-house, in the design offices of large aircraft and car manufacturers. All the main companies in these two sectors ran their own software development groups. These included the car firms Ford, GM, Mercedes-Benz, Renault, Citroën, Nissan, Toyota, and the major aerospace firms Lockheed, McDonnell-Douglas, Northrop, Boeing, British Aircraft Corporation and Dassault. These organisations dominated the world of CAD mainly because they were the only ones who could afford the only computers available at the time: large, very expensive, mainframe machines.

The principal use of these systems, at this time, was in two-dimensional drafting. However, the firms and researchers in the area were actually pursuing a much more sophisticated idea of computer-aided design. This was how to design products in such a way that the resultant design information could be passed directly to

---

10 Schodek et al., p 5.

numerically controlled (NC) tools and other types of computerised machines, for manufacture. The main intellectual challenges were how to represent complex surfaces and solid objects in mathematical terms, so that they could be modelled in computers.

The first problem – the mathematical modelling of complex surfaces – was probably solved initially in about 1958, by a mathematician called Paul de Casteljau, working for Citroën. Citroën kept his work secret, but it was duplicated soon after by Pierre Bézier who was working for Renault at the time. As a result of Citroën's secrecy, it is Bézier, rather than de Casteljau, whose name is associated with the mathematical approach to curves and surfaces that is most widely used in CAD systems today.[11]

The theory of surface modelling is complicated, beyond the scope of this book and certainly beyond the mathematical capability of the writer. However, like much of today's technological world, one does not need to understand how things like non-uniform rational basis splines (NURBS) really work, to be able to use them more or less intelligently in a drawing program. Nonetheless, there are some important components of this work that are worth dwelling on, briefly. The first is the idea of splines.

Splines come from boat building. The traditional way of building the hull of a boat starts with a scale drawing called a lines drawing, and what is called a table of offsets. The lines drawing comprises a plan view, a long section, and a pair of half cross-sections of the vessel, one looking forward, the other aft. Each of these views shows the outline of the hull together with key contour lines which trace points on the surface that are constant distances either from the hull centre line or from the design waterline of the boat. The table of offsets gives the precise location, relative to the waterline and to the hull centre line, where each of the contours cross a set of transverse cross-sections spaced equally along the length of the boat. These points are called stations.

The objective in the process called lofting is to take the lines drawing information and scale it up to full size, either on a 1:1 scale drawing, or directly onto the sheet material from which the hull will be built, so that the plywood, steel, or other material can be cut accurately. To do this the overall rectangular grid of the hull is first drawn out precisely. The contours are added to the drawing or hull sheeting by fixing pins, or blocks called ducks, at the points on the full-scale drawing, or the material sheet, corresponding exactly to the locations of the stations given in the table of offsets. These station points are then joined up smoothly, so as to give a fair contour line on the hull. This is done by threading a strong flexible strip of wood or metal, called a spline, along the sequence of pins. The spline comes to rest in a form that is made up of shapes of minimum strain energy between each of the station points. The contour is drawn tracing the line that the spline follows onto the underlying material. This creates a smooth interpolation of points between successive stations on the contour line. The result is an approximation to the

---

11 Bézier, P., 'A View of the CAD/CAM Development Period', *IEEE Annals of the History of Computing,* 20(2): 39, 1998.

original contour on the lines drawing, close enough to the original to generate a fair line, but requiring far less calculation and measurement than a true copy would require.

The spline concept is used to serve a similar function of simplification and approximation in computerised drawing systems. In order to plot a curve on screen or on paper, the computer must solve the equation of that curve for every point along its length. This is not a serious problem with simple curves such as straight lines and conic sections, which can be described explicitly by straightforward formulae like $y = ax^2 + bx + c$. However, the curves which describe the surfaces of ships, airplanes, cars and many household products for example, are almost uncomputably complex, with many changes of direction and curvature. Splines in CAD systems approximate these curves by breaking each of them down into smaller, relatively simple, separate sections, solving the equations for those sections, and then re-joining the sections through a process of numerical trial and error until a smooth, continuous connection is achieved. Old-fashioned hand-drafting used French curves to achieve a similar effect.

The development of the splines concept as a means of simplifying and approximating complex curves and surfaces was a crucial step in the history of CAD. Following de Casteljau and Bézier, most of the innovative work in this area was carried out by a relatively small number of academics, mainly in the USA and at Cambridge, notably Steve Koons at MIT and Robin Forrest at Cambridge.

The data from these surface modelling systems can be passed to a variety of capable of milling or pressing large complex curved surfaces. The second big problem for manufacturers was how to model solid objects like engine blocks, and assemblies of objects like machine guns. Perhaps because solid modelling is more conceptually difficult or computationally demanding, it was not until the 1970s that the first solid modelling systems were released. There are many different approaches to the problem of modelling a solid object; the two most widely used techniques are constructive solid geometry (CSG), and boundary representation (BREP). A CSG model is created by performing Boolean set operations – union, intersection and difference – on primitive solid shapes, such as cones, pyramids, cylinders and cuboids. BREP, as the name suggests, works by tracing and recording the edges or boundaries and vertices of solid forms. CAD systems typically combine elements from both of these.

One of the first solid modelling programs, called Part and Assembly Description Language (PADL-1), was created by Herb Voelcker at the University of Rochester, and was released in 1978. PADL was written in Fortran and so had the advantage that it could run on many makes of computer. The program and its successors were widely used in US universities and industry until the late 1980s.

Most modern CAD systems however, trace their origins back to a program called BUILD-1, created by Ian Braid at Cambridge University and also first released in 1978. Braid, a colleague of his, Charles Lang, together with a few others at Cambridge went on to develop the ACIS and later Parasolids modelling programs. Between them, ACIS and Parasolids provided the modelling kernels for the great majority of subsequent CAD systems.

5.3.4 The early CAD industry

The structure of the CAD industry, as a discrete industry sector, began to emerge from the mid-1960s onwards as the underlying hardware platforms developed. Throughout the 1960s, most CAD development work was carried out in universities or by in-house programming teams in major manufacturing companies, and all were working on mainframe computers – notably IBM machines. All of these user organisations developed their own CAD software programs.

The only real survivors from this type of software are CADAM (Computer-graphics Augmented Design and Manufacturing) created by the Lockheed Corporation and CATIA (Computer-Aided Three-dimensional Interactive Application), which was developed by Dassault on the basis of Bézier's work at Renault and a CADAM source-code licence purchased from Lockheed. The functional ambitions of both these products – linking computer-aided design directly with manufacturing processes, and doing real 3D modelling, is nicely reflected in their acronymic names.

Mini-computers first appeared in the late 1960s and were in active use as CAD platforms by the early 1970s. The most widely used minis were those of Digital Equipment Corporation (DEC) and Data General (DG). Systems were still phenomenally expensive. A typical single user station cost about $150,000 in 1972, additional stations cost $50,000 each – about a million dollars in today's money. For this the customer got a 16-bit computer with 8K–16K of memory, a 10Mb – 20Mb disc drive, a digitiser pad, a plotter and a few other bits and pieces.[12] A decade later the seat price was about $130,000, but the buyer got a much bigger bang for his buck.

The CAD industry saw its first period of rapid growth during the 1970s. This was based partly on the improvement in hardware economics, but mainly on the fact that the new software concepts, surface and solid modelling specifically, offered major manufacturing customers unprecedented opportunities for improvement in their CAD/CAM processes. At this time CAD was sold almost exclusively on a turnkey basis, in which companies that were primarily hardware vendors provided a complete computing bundle, including their own CAD software built for their particular machines, but generally incorporating licensed kernel software such as ACIS and Parasolids. The main companies in this group were Applicon, Computervision, Auto-trol Technology, Calma and Intergraph. The CAD software and hardware market grew from under $25m in 1970 to just under $1bn in 1979.[13]

The 1980s saw a new phase in the development of the CAD industry with the introduction of Unix workstations by firms like Apollo, Sun Microsystems and Silicon Graphics. In that they didn't need special air-conditioned rooms and ran

---

12 Weisberg, Section 2, p. 9.
13 CADAZZ http://www.cadazz.com/cad-software-history-1970s.htm (retrieved 18 June 2010).

(almost) generic Unix, some with graphical user interfaces, these really changed the way in which engineers used computers. They were also far cheaper than the earlier mini-computers, so could often be bought out of departmental budgets, rather than through central IT departments.

(A final CAD package to note here is Plant Design Management System (PDMS), a 3D process plant modelling system designed initially by Dick and Martin Newell of the CADCentre in Cambridge and first released in the 1970s. CADCentre, later renamed Aveva, and Intergraph now dominate the market for design, asset management and facilities management in the major plant, offshore and shipbuilding industries.)

5.3.5 CAD and the personal computer

However, the IBM Personal Computer and its DOS operating system, released first in 1981, completely disrupted the young, workstation-based market. John Walker, the driving force in the group of systems people who created Autodesk, wrote at the time '(It was) … in December of 1981 that I first formed the idea of starting a software only company to provide software for the coming tidal wave of small computers …'. His ambition was to create 'the next Visi-Calc', as he put it. The group already had a piece of software called Interact, which in Walker's words was 'a superb product in a virgin market'.[14] It was released as AutoCAD in 1982.

It is a tribute to Walker's vision and charismatic leadership that Autodesk survived and prospered in the early chaotic days of PC CAD. It might be unreasonable to extract one quotation from his book as being representative of his strategy, but the following comes very close: 'Autodesk has always competed like a hungry rat. We will continue. And we will prevail.'[15] The book is a remarkably good read; insightful and inspirational. As an account of the relatively early days of IT, it's in the same league as Tracy Kidder's *The Soul of a New Machine*.[16]

Bentley Systems' MicroStation PC followed in 1985. MicroStation was created as a PC version of Intergraph's Interactive Graphics Design System (IGDS). Throughout its first decade of existence, MicroStation was largely seen as an Intergraph product, part of the Intergraph organisation, and marketed by Intergraph, with the Bentley team acting as a development group within the larger company. So in a sense, throughout the crucial first ten years of brand development, Bentley was protected from the PC competitive jungle in which Autodesk gradually became dominant. And starting from IGDS, a genuinely sophisticated CAD product, Bentley's MicroStation was inherently more powerful, more complete as a design tool, than AutoCad, which was created specifically as a drafting package. This and the fact that it was available for Apple Mac machines, made MicroStation particularly attractive to some of the more technologically ambitious design firms in the AEC industry.

---

14 Walker, J., *The Autodesk File:Bits of History, Words of Experience*. pp. 13, 17, 39 at http://www.fourmilab.ch/autofile/ 1994 (retrieved 25 June 2010).
15 Ibid. p. 407.
16 Kidder, J.T., *The Soul of a New Machine*. New York: Atlantic-Little, Brown, 1981.

However, by the mid-1990s, when the Bentley team split from Intergraph, AutoCad was pretty well comparable in functionality with MicroStation, except perhaps in the areas of road/rail and process design. And today, whatever their comparative merits, these two are the main survivors of the early, deadly frenzy of competition for the PC CAD market.

5.3.6 *The early days of AEC CAD in the UK*

The great thing about the PC (and Mac) computer-aided drafting systems was that they made it possible for even the smallest companies to produce professional looking drawings. One might say that AutoCad democratised CAD; brought CAD to the people. It remains the case that, although the mainstream systems have developed some more advanced design capabilities over time, the vast majority of even the largest user companies employ these systems more or less entirely for 2D drawing production. The surviving systems have done just what their users mainly wanted them to do – produce drawings efficiently – and they have been enormously successful as a result.

Which, in a sense, is a pity. Because, in the UK, since back in the late 1970s and early 1980s, somewhat on the periphery of early mainstream CAD, a number of firms and individuals had been trying to get true, model-based, architectural CAD off the ground. Acropolis CAD, developed by Building Design Partnership, Gintran from Bristol University, and GDS by Oxford Regional Health Authority are three of these. A fourth and the most enduring was RUCAPS, developed initially by John Watts and John Davidson at Liverpool University and subsequently taken up by GMW, one of the largest architectural practices in the UK at the time. All of these systems were based on the idea of a three-dimensional, component-based model from which drawings and other data could be derived – they were not conceived of as being just drafting systems, but were true modelling systems from the beginning.

The RUCAPS team at GMW included Jonathan Ingram and Robert Aish, of whom more later. RUCAPS ran on mini-computers; it was expensive and complicated. It died away, but the ideas lived on. Ingram and a small team next created a workstation-based system, called Sonata, a modestly successful product of the early 1990s. Sonata disappeared in a mysterious, corporate black hole, somewhere in eastern Canada in 1992. But Ingram went on to create a further generation of the family, called Reflex, in the mid-1990s. Reflex was subsequently bought out by Parametric Technology Corporation (PTC) in 1996.

Stories converge at this point because, back in the mainstream CAD world – generally referred to as Mechanical CAD (M-CAD) – PTC had caused major disruption when it brought its ProENGINEER product to the market in 1987. Other systems, in one way or another, possessed most of the features boasted by PTC, but ProENGINEER brought them all together in a very high performance package that was marketed very aggressively.

For present purposes, the key features of ProENGINEER are the fact that it is inherently solid model based, all its design and analysis functions use the same database and data structure, objects are parametric (see below) and the system manages design development and change very effectively, by saving what is called a history tree. This keeps track of every action carried out in the creation of the model, as they are made, and allows them to be reversed out relatively easily.

In the course of its rapid expansion, through the 1980s and early 1990s, PTC moved into almost all of the industries where computers were used in design and manufacturing. AEC was an obvious target, but a difficult one. However, Reflex seemed to offer a route into AEC that was complementary with PTC's parametric modelling approach. So PTC bought Reflex and released PRO/Reflex in late 1996. Unfortunately, the fit was not as close as PTC had initially believed, and AEC was a more difficult market than had been expected, even for such intrepid marketeers as PTC. The company sold the product to the US firm, Beck Construction in 1997. Beck took Reflex and used it to create a parametric estimating package which they use in early-stage analysis of their design/build development projects.

Also in 1997, independently of the sale to Beck, a group of people from PTC who had worked on the Reflex project, set up a new company called Revit Technology Corporation, selling a product that offered a single-database, parametric component-based design system with a remarkable history management and change propagation capability. The Revit name presumably reflects the fact that 'Revit's parametric change engine automatically reflects any design change throughout the entire project, managing all CAD chores related to the project while the architect concentrates on the design intent', as a company press release of early 2002 suggests.[17] Autodesk bought Revit later that year and the BIM snowball started to roll.

5.3.7 Parametric modelling

The key feature of BIM systems is the ability to create and manipulate three-dimensional building models using highly accurate, parametric objects to represent the physical components of real buildings. A parametric object is one that belongs to a class or family of things that all share certain properties. One can think of there being two types of properties: fixed properties that all members of the same family share; and variable, or parametric, properties, which distinguish individual family members from each other.

Figure 5.1 illustrates this concept. Shape A and Shape B are both members of a family of cam-shaped steel plates with a threaded hole in one end, into which a bolt is to be screwed. The shared, or fixed properties of all members of this family might include the fact that they are all made from a particular grade of steel and plate thickness. The variable, or parametric features shown here are: the two end shape radii R1 and R3; the hole radius R2; and the centre to centre dimension L1. These parameters vary from member to member within the family. Figure 5.1

---

17 http://www2.prnewswire.com/cgi-bin/stories.pl?ACCT=104&STORY=/www/story/01-10-2002/0001645545&EDATE (retrieved 28 May 2010).

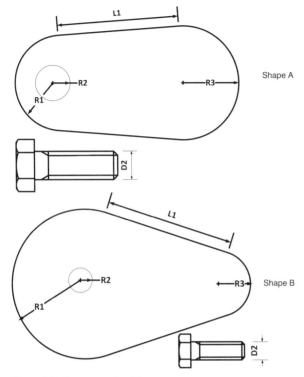

Figure 5.1 A parametric object

shows two plates whose L1 values are equal, but R1, R2 and R3 values are all different between the two shapes.

To design an assembly using this type of parametric component, the user simply selects the family of components he needs from a menu, and then adjusts the various parameters associated with that component family to arrive at the specific required object.

The ability of objects to display systematic behaviour is a key concept in parametric modelling. It means simply that whenever one characteristic of an object is changed by the user, other related characteristics also change in some predictable, programmed way. This can include properties such as the orientation of one component relative to another or to a specific set of coordinates. For example: component A is defined as being a certain distance from, and parallel to, component B, so whenever component B is moved, component A also moves, to maintain the parametrically specified relationship. The direction of causation is obviously important here. In the example shown above, the assembly combines the plate and a bolt that passes through it. The value of the hole radius R2 might be driven by the diameter of the bolt: $R2 = D2/2$, for example, but it would be equally possible for the bolt diameter to drive the radius of the hole, expressed perhaps as $D2 = 2 \times R2$.

As even this very simple example indicates, the use of parametric components in the design of complicated assemblies and products can increase the designer's productivity dramatically. However, the real beauty of parametric methods lies in their ability to manage component properties other than geometry. For example, in Figure 5.1, the difference between R1, the end shape radius and R2, the hole radius might be limited so as to ensure a safe thickness of material at that end of the plate. This can be achieved by inserting into the parametric definition of the cam, a simple expression like R1–R2 ≥ X, where X is some permissible minimum thickness of material.

This little example is interesting for two reasons. First it illustrates the general ability to specify components in a richer sense than basic geometry, provided only that those additional properties can be expressed parametrically. So, attributes such as density, u-value, modulus of elasticity and so on can all be applied parametrically. Also, of course, 'economic' properties such as unit cost, planned and actual delivery dates, construction man-hours and so on can all be built in to the specification of components in a parametric building model.

Crucially however, knowledge or intelligence can also be programmed into these components. In the example given, the engineering calculation that a minimum thickness of material X must be allowed at the end of the cam piece, is programmed into the behaviour of the component. This is a one-time exercise; once that particular piece of engineering knowledge is embedded in the definition of that particular family of components, it remains and the design never has to be re-addressed.

Thus, the real sense in which parametric modelling systems are said to use 'intelligent' components is because these components both behave predictably and encode potentially unlimited quantities of human knowledge. These capabilities have enormous implications. As Weisberg puts it:

> In very simple terms, virtually no product, building, electronic component or system or factory is designed today in a developed country without the use of this technology. It has resulted in more reliable products that are less expensive to produce and are more attractive to potential customers. It has changed technical education and to a significant extent, the practice of numerous professions.[18]

5.3.8 European developments

In the English-speaking construction industry, Archicad is something of an overlooked BIM pioneer. The system, which was created by a Hungarian company called Graphisoft and first released in 1984, was one of the first CAD packages developed for the Apple Macintosh. It was the first architectural CAD system to provide true 3D modelling capability. It was also the first personal computer-based system to enable the user to associate information other than geometry

---

18 Weisberg, Section 2, p.21, 'The Engineering Design Revolution'.

with objects in the model. As such Archicad was the first to attempt to integrate architecture, engineering and construction information in a single intelligent model, an approach the company referred to as virtual building, or Virtual Construction (VICO) as it later became.

A second notable European CAD system was Allplan, by Nemetschek AG, a German civil and structural engineering firm company founded in Munich in 1963. Allplan, which was based on Nemetschek's earlier work on computer-aided engineering and analysis systems, was first released in 1984. Like Archicad, it was also developed initially for the Apple Mac and was designed as a 3D modelling and engineering system, rather than as a drafting tool. Nemetschek acquired Graphisoft in 2007, at which point the VICO operation was spun out as a separate company.

Diehl Graphsoft developed MiniCad, later re-named Vectorworks, also a 3D modelling package for the Mac platform. The company was acquired by Nemetschek in 2000 and renamed Nemetschek US.

It is sadly significant that two such technically advanced, (inexpensive) and competent products as Archicad and Allplan should have struggled to establish and maintain customer support in the English-speaking construction industry. It may be the case that they have simply been out-marketed by firms like Autodesk and Bentley, particularly in North America.

But it is more likely that they were victims of the process described by Kalay as 'the "dumbing down" of architectural CAD, (which) happened while other disciplines were making their own CAD software more intelligent'.[19] There would seem to be little prospect of success for intelligent design software in an environment in which as Kolarevic puts it, 'The standard contracts in use by the AIA state explicitly that: "the architect will not have control over or charge of and will not be responsible for construction means, methods, techniques, sequences or procedures." '[20] In this context, where the architect must provide only the barest of 'design intent' information, design firms are forced to focus on efficient drawing production above all else.

5.3.9 Recent events; current state of play

By the beginning of the present century, ProENGINEER dominated the mechanical CAD industry. Of the large systems that originated in the mainframe era, only a few, including UGS and SDRC, both now rolled into Siemens' PLM system and Dassault's CATIA, have managed to stay the course. The others, and the hardware-based mid-range and workstation vendors like Intergraph, were simply outcompeted. There has been remarkably little recent innovation in the core capabilities of CAD systems in the mechanical CAD arena in recent years.

---

19 Kalay, Y.E., *Architecture's New Media. Principles, Theories, and Methods of Computer-Aided Design.* Cambridge, MA: MIT Press, 2004. p.71.
20 Kolarevic, p. 58.

The major problem remaining is applying the technology to increasingly complex projects. That means managing massive amounts of design data – a task some companies are doing well while others are struggling.[21]

Thus, the big players have tended to focus on product lifecycle management (PLM), vendor-speak for systems needed to solve the problems associated with the enormous quantities of documents generated by people using CAD systems. PLM systems combine aspects of CAD modelling with document management, facilities management and geographical information systems, to enable the manufacturers of complex products and the owners of major facilities to manage their assets throughout their life-cycles. In some cases these efforts extend to integration or other linkages with corporate enterprise resource planning (ERP) systems. The mainstream vendors are also trying to address the challenge of collaborative working, concurrent engineering, model sharing, or whatever, generally over the internet, and generally using web browsers.

While the AEC vendors have also, of necessity, been pushing their versions of PLM and collaborative working, in contrast to the relative stasis in mechanical CAD, there have been some striking developments in the AEC marketplace in recent years.

One of the most visible and most dramatic of these has been the collaboration between Frank Gehry's architecture firm and Dassault Systèmes, who supply the CATIA system, used by Gehry to design and construct his remarkable, impossible buildings. Gehry's practice, initially under the technical direction of partner Jim Glymph, has been using CATIA in-house since their Barcelona Fish project of 1992. Gehry's unique methodology, which works by integrating information flows between designer and component manufacturers, has been deployed on many of the most remarkable buildings of the past decade, notably perhaps the Guggenheim Museum in Bilbao.

The firm has recently set up an offshoot called Gehry Technologies which has entered into a partnership with Dassault to develop and market a new version of CATIA, called Digital Project (SP). This is CATIA customised for use in architectural design and construction. Digital Project is a relatively expensive system, but it has developed a significant following amongst the designers of complex buildings. For example, *AEC Magazine* reports a sale of 100 DP seats to SOM in June 2007.[22]

A second interesting area of activity is concerned with the design and construction of 'blobby' architecture;[23] buildings and other structures that are organic looking, multiply curved and complicated, both as surfaces and as usable spaces. The Smart Geometry (SG) group is probably the main forum of activity in this space. SG includes academics, practising designers from such firms as Foster & Partners, Arup and Buro Happold, and software designers, all collaborating in

---

21 Weisberg, pp. 2–22.
22 http://aecmag.com/index.php?option=com_content&task=view&id=171&Itemid=37 (retrieved 20 May 2010).
23 Kolarevic, p. 57.

The origins of BIM in computer-aided design 77

a manner reminiscent of the way in which the earliest pioneers of computer-aided design shared and developed ideas in the 1950s and 1960s.

There are two particularly interesting aspects of this work. The first is the pursuit of algorithmic and other mathematical techniques as a way of describing and exploring complex forms. The second aspect is to do with the process by which the software is being developed – as a very close collaboration between building designers and software engineers. Blobby buildings will probably remain a minority taste for the foreseeable future, but the architectural design and software development techniques being explored by these researchers is potentially very significant. To quote Robert Aish, formerly of Bentley, one of the leading lights of this community:

> The challenge that the software developers should be keen to accept is how to create a new type of design tool that can respond to the opportunities presented by these new, more exploratory approaches to architectural design.[24]

This sort of open, collaborative, non-proprietary research is particularly important in the early 'blue-sky' stages of development of design, or mathematical or systems concepts. As David Weisberg described the early work on surface modelling:

> Much of the work going on in developing better surface definition techniques was being done at academic research centers and was typically published in widely available journals. Each researcher was, therefore, able to build on the work of those who had tackled earlier aspects of the problem. As seen by what occurred at Citroën, this would probably not have occurred if the work had primarily been done by industrial companies.[25]

Bentley MicroStation, with its sophisticated scripting capability is the main software platform used in this work. Bentley have recently released a new product called Generative Components, intended specifically to support designers adopting this approach.

The Swiss Re 'Gherkin' building, at 30 St Mary Axe, London, involved an interesting combination of physical modelling – 'analogue building', as Robin Partington, Foster & Partners director on the project, put it[26] and 'blobmeisters' from his company, Arup and elsewhere.[27] The resulting building is a sophisticated hybrid whose external skin includes large components that are machined to a tolerance of 0.1 mm, a truly fantastic degree of precision by construction industry standards.[28]

---

24 Aish, Robert, 'Extensible Computational Design Tools for Exploratory Architecture', in Kolarevic, p.245.
25 Weisberg, pp. 2–12.
26 Powell, K., *30 St Mary Axe, A Tower for London*. London: Merrell, 2006, p. 210.
27 Ibid., p. 63.
28 Ibid., p. 84.

Autodesk, meanwhile, has hardly been standing still. The company greatly expanded its range of products with the introduction of 3D Studio Max for 3D modelling and animation in 1997 and Inventor, a solid modelling application, in 2000. However, Autodesk has made five particularly notable strategic moves in the AEC marketplace in the past eight years. The first was the purchase of Revit in 2002, as noted above.

The second, five years later, was the purchase of Sheffield-based Navisworks, a multiple file aggregator and viewer, a very useful sort of BIM-lite product that enables the user to create models using multiple CAD files, of different formats, from a variety of sources, which can be used for fly-throughs, operations simulation and clash checking.

Autodesk's third key initiative was the recruitment of Robert Aish from Bentley in January 2008. Although nothing has yet emerged as a result of this move, it is to be expected that he will bring some of the early-stage, form-finding capabilities of Generative Components, and some of the software design techniques of the Smart Geometry group to the Autodesk armoury.

The fourth key initiative was the launch of the Autodesk Seek service in May 2008. Seek is a repository of digital building product information, similar in a sense to the Barbour Index service in the UK. But whereas Barbour is essentially a repository of product catalogues, containing product specifications and non-editable illustrations, Seek also provides access to manufacturers' product CAD files and BIM model files. This is obviously a complicated area of development, with many issues of intellectual property and commercial confidentiality to be resolved, in the UK at least. However, providing free public access to editable, reusable versions of their product data seems to be working for growing numbers of US manufacturers.

> Autodesk ... announced that the company's Autodesk Seek web service is averaging more than 900,000 searches per month, while providing nearly 300,000 downloads of building product information and models to architects, engineers and other design professionals per month. This represents a year-over-year increase of over 165 percent.[29]

McGraw-Hill Construction, whose Sweets division is the major source of product catalogues and estimators' pricing books in the United States, has recently announced a collaboration with Autodesk in this area. All Sweets' manufacturers are represented in Seek, and vice versa. And the two companies go to significant lengths both to help manufacturers get involved and to make specifiers aware of the ease with which product information can be assessed.

(Behind the surface of easy-to-get-at information, Seek is actually a very interesting and ambitious project in taxonomy – the area of study that deals with the problem of identifying and classifying the things that are known to exist in a particular domain. In this case the domain of interest is construction. Any reader

---

29 Autodesk press release, Thursday June 10, 2010.

who is interested in the seemingly perpetual mess called construction industry classification is encouraged to view a talk given by a senior Autodesk developer, Mike Haley in July 2010.[30])

Autodesk's fifth big, interesting move was the announcement of a format sharing agreement with Bentley Systems, also in 2008. This is intended to enable 3D information and some component intelligence to be exchanged between the two sets of BIM products. It's not complete interoperability, but it should help greatly to simplify this long-standing and messy problem. As the technology commentator Martyn Day put it:

> This agreement is really co-opetition, making life easier for everyone all-round. It's the most significant and positive move in the history of Computer Aided Design in the AEC space. If the agreement continues and both parties play fair then there will be significant advantages in the coming years for the industry as a whole ... Conjecture aside. It's a time to rejoice.[31]

5.3.10 Conclusion

The main purpose of this chapter has been to convey a sense of just how difficult, or how technically challenging, computer-aided design is – as a form of computing. CAD is a remarkably complex activity that places huge demands on even the most advanced hardware and software technologies. There are almost no other areas of commercial computing that require such an intense combination of high-performance database management and powerful graphics processing capabilities. And, in a sense, building design is one of the most challenging types of CAD. As noted in Section 5.1.1, a key characteristic of AEC CAD is that building models are required to handle exceptionally large numbers of components. The overall geometry of buildings is usually relatively simple, but the components that make them up can be as geometrically complex as any other manufactured products. The conjunction of this with the many possible ways in which building components can be assembled, connected and combined with each other, presents a fantastic computing challenge. So even basic AEC CAD is difficult to do well; advanced CAD – Building Information Modelling – is, and will continue to be, a real challenge.

There are a couple of observations worth noting with regard to the development of CAD concepts and technologies over the past 50 years. First, most of the histories of CAD currently available – almost all of which are Web narratives – tend to dwell on the American contribution. It seems important to balance this picture by drawing attention to the fundamental work of the Frenchmen, de Casteljau and Bézier, and also to note the remarkable contribution of British researchers, Lang, Braid, Forrester and others on surface and solid modelling, the Newell brothers on plant modelling, as well as the work on architectural modelling by Watts and

---

30 http://fora.tv/2008/03/19/Mike_Haley_on_Autodesk_Content_Search (Retrieved 1 June 2010).
31 Day, M., 'Comment', *AEC Magazine*, July/August 2008, Vol.38, p. 9.

Davidson and their successors, notably Ingram and his team, without whom the evolution from RUCAPS to Revit might never have happened.

The second observation to note here is how many of the fundamental intellectual breakthroughs originated amongst a relatively small number of mainly academic groups, working in different parts of the world. There is no question that the financial support of industry, the military and other branches of government was essential to the success of these groups. But nor is it likely that the concepts would have been exploited as well as they were without the vision and entrepreneurship of a relatively small number of remarkable business people.

There is no sense in any of this of there being a guiding hand at work; there was no grand design. But the history of computer-aided design is the story of one of the great intellectual and commercial achievements of the twentieth century.

6 Building Information Modelling

6.0 Introduction

The central proposition of this book is that one of the main causes of underperformance in the construction industry is poor-quality design information and ineffective communications amongst the members of project teams. The BIM approach promises a solution to both of these problems. It generates dramatically higher quality design information, and also enables that information to be managed and communicated far more efficiently than in the past.

The diagram in Figure 6.1 illustrates, in a very high level sense, the main types of information generated on a typical construction project. The diagram centres on the activities of architecture and project management and the constellation of other business functions, disciplines and trades that surround them. The great hope for BIM is that it will enable all of the players in all of these functional areas both to generate truly high-quality information and to be able to exchange that information effectively and efficiently with other people in the diagram.

Until fairly recently, there was a view that a BIM model could actually take the form of a single unitary model, sitting on a single database, accessible simultaneously to all members of the project team. That view has since moderated. It is now generally accepted that the project model is more likely to take the form of a federation of separate, but interconnected, discipline-specific sub-models. This chapter considers how this can be done.

The definition of BIM used in this book is as follows.

Building Information Modelling is an approach to building design and construction in which:

- A reference model of the building is created using one or more parametric component-based, 3D modelling systems.
- These systems exchange information about the building in one or more agreed standard file formats, with each other and with other systems which conform to the agreed formats.
- These exchanges are regulated by a set of protocols which establish the particular types of information to be exchanged between different members of the team, at different points in the project life-cycle.

82 Building Information Modelling

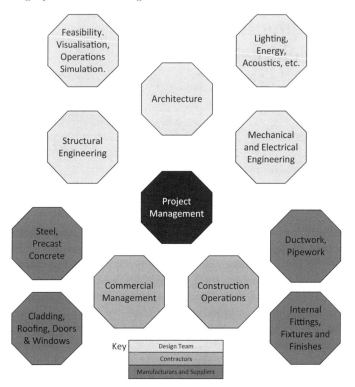

Figure 6.1 Information generated on a construction project

This idea of BIM involves a three-layered approach: the BIM reference model, standard formats for data exchange, and information interchange protocols (Figure 6.2).

This approach to BIM is generally consistent with that of Chuck Eastman and his co-authors: '... a modelling technology and associated set of processes to produce, communicate and analyse building models'.[1] The purpose of the present re-phrasing is to expose the three aspects of BIM more clearly, so that they can be addressed explicitly in the sections that follow.

6.1 BIM authoring tools – characteristics of BIM systems

Chapter 5 described the development of computer-aided design (CAD) technology up to the point where Building Information Modelling appeared on the scene, in the early years of the last decade. This chapter takes up the story from that point. Recall that CAD originated with basic 2D drawing tools, back in the 1960s.

---

1 Eastman, C.M., Teicholtz, P., Sacks, R. and Liston, K., *BIM Handbook A Guide to Building Information Modelling for Owners, Managers, Designers, Engineers and Contractors*. Hoboken, NJ, John Wiley & Sons, 2008, p. 13.

Building Information Modelling 83

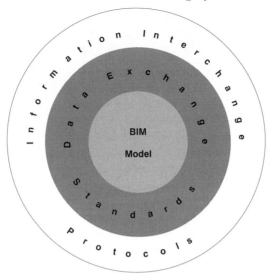

Figure 6.2 BIM – a three-layer approach

In some environments a sort of pseudo-3D effect was achieved using wireframe models, sometimes with surface shading to add a solid appearance. Methods of modelling complex surfaces, such as the bodies of cars and aircraft, followed and true solid modelling systems first appeared in the mid-1980s. Parametric methods for assigning and managing the properties of model components also became mainstream at about that time, although the use of parametric techniques was limited largely to mechanical CAD systems – used mainly in manufacturing industries.

The world of AEC CAD remains frozen in the earliest stage of this technology evolution. Only recently, in its tentative explorations of the BIM idea, has construction shown any inclination to move on from its dependence on basic drafting techniques and associated computer systems. There is nothing illogical or particularly reactionary about this; as long as drawings remain the preferred way of designing buildings, drafting systems will, correctly, remain the technology of choice.

However, model-based design is different to drawing-based design, and BIM modelling systems are very different to drafting systems. The definition of BIM in Section 6.0 is based on the idea that the building design is created using parametric component-based modelling systems. The key features of parametric modelling systems are, briefly, as follows.

They work by enabling the user to define a virtual 3D space and to insert into that space, objects that correspond directly with the components of buildings as they exist in the physical world: walls, doors, pumps, tanks and so on. The systems store and organise these virtual components in families or classes of objects that share certain properties. For example, walls are vertical structures that separate adjacent spaces from each other. The properties that walls share include length, height and thickness. That is to say, every wall, by definition, and in order to be classed as a wall, must have a length, a height and a thickness, otherwise it cannot be a wall, but must belong to some other class of object.

These properties of the wall component are said to be parametric, in the sense that they each have a possible range of values, a specific one of which is chosen when the user selects a particular wall. Thus, a class of objects called walls can be defined, all of whose members must have a length, whose value can range from, say 0.5 m to 100 m. When the user inserts a particular wall into a model, one of the properties he selects is its length, say 10.00 m. He must also, of course select, or fix in the same way, the values for the thickness and height of his chosen wall.

Length, height and thickness are obvious properties of walls in the context of conventional drawing systems. The crucial feature of BIM systems is that they enable a much broader, more extensive range of properties to be associated with the objects they support. These properties can include:

6.1.1 Physical attributes

Properties, such as density, u-value, compressive strength, and so on, which are properties of walls generally, can be used as the basis for selecting a particular wall. The value of each of the properties associated with a particular wall is stored in the BIM database and so can be retrieved and used for scheduling, calculation and analysis, by the BIM software itself, or in some cases, by other applications.

6.1.2 Economic attributes

These can include properties such as the unit cost of an item, manufacturer details, delivery lead time, estimated erection man-hours, special handling requirements and so on, all of which can be used internally within the BIM application, but which can also be shared with other interfacing applications.

6.1.3 Relationships

Objects can be related to other objects. For example, it is possible to stipulate that a particular wall must be parallel to and a specified distance from another wall; that it is attached to a third wall at a particular angle, that it is perpendicular to the floor it rests on, and so on.

6.1.4 Behaviour

Two or more parametric properties of a given object can be specified as being related in some systematic or mathematical way; when one parameter is changed, the value of the other changes according to the formula. For example, the length of the wall can be specified as being equal to five times its height. If the height is subsequently increased, the length will automatically be increased as specified in the formula. There are two things to note. First, these relationships can be very rich, but can become complicated very quickly. Second, parametric relationships can span across objects. So, for instance, the length of the parallel wall in our example can be specified as also being equal to five times the height of the first wall, so both walls increase in height when the first one is stretched lengthwise.

6.1.5 Intelligence

An extension of the idea of behaviour is that higher level rules can be embodied in objects using parametric attributes. For example, BIM systems allow windows to be inserted into walls. A rule might say that a window may not be inserted closer than 0.25 m from the end of a wall. Any attempt to insert a window closer to the end than this generates an alert.

6.1.6 Self-awareness

Objects 'know' the space they occupy and can be set to generate alerts if any other object is inserted into or otherwise impinges on that space. This enables automatic clash detection and avoidance.

6.1.7 Spaces

A building is made up of physical components. These shape the spaces – areas and volumes – in which the building's functions are carried out. BIM systems know about spaces explicitly, as objects, just like components. The spaces within a model can thus be used like components in that they can be manipulated, scheduled, analysed and so on.

6.1.8 Connections

An extension of the idea of intelligence is that objects can be programmed to know about other objects to which they can be connected, what form of connections can be used for this and how the connection should be constructed. A simple example is how two pipe spools can be jointed. A more complex one might be the detail of how an external, double-skin brick wall sits on the edge of a floor slab.

The BIM modelling tools currently available all provide extensive libraries of off-the-shelf, standard, well-specified objects, and they provide methods whereby users can extend the range of properties associated with their standard object families. They also enable users to create their own custom families, and individual family members, and to do so in such a way as to ensure that the new families and individual components behave correctly within the system.

In addition to the generic components that are provided by the BIM tool vendors, there are two types of proprietary component families available. The first of these is components representing the products of construction product manufacturers. These are created by the manufacturers, or by third party specialists on their behalf, and embed specific features, different or additional to, the BIM vendors' standard, generic components. The second type of proprietary components are those created by independent software companies and sold on the open market. Both of these types of families are provided by their originators, as downloadable files in a variety of formats, usually one or more BIM vendor formats and usually in the neutral IFC format. (See Section 6.3 for discussion of data formats.)

Autodesk, as noted in Chapter 5, has introduced a component repository service called Seek which enables manufacturers to upload their product models to the web. These can easily be retrieved by designers and incorporated into their work. Competing services, such as Barbour Index, can be expected to move into this area in the foreseeable future. So there is an active community, comprising BIM vendors, independent developers and construction product manufacturers, all at work creating new, well-specified, generic and proprietary component models for the major BIM authoring systems.

6.1.9 Implicit knowledge and embedded knowledge

In conventional – let's say old-fashioned – architectural design, a compact existed between the architect and the artisan who built his ideas. One the one hand the architect undertook to know enough about the materials the artisan worked with, and about the tools and techniques he employed to be able to direct him easily and unambiguously in his work. The artisan for his part undertook to know enough about the design process to be able to read the architect's drawings and to understand and be guided by them, easily and unambiguously. They both spent a significant portion of their training – articles or apprenticeship – learning these respective skills, and a huge volume of knowledge was implicit in their dialogue. That's the essence of craftsmanship.

Twenty-first-century construction has no place for such practice. Modern buildings comprise such a vast range of specialist equipment and materials that no amount of article-ship would enable a trainee architect to master their handling and erection. Hence the importance of 'design intent'. In this mode of practice, the architect nominates a product from a catalogue and effectively issues sketches indicating the sort of result he wishes to achieve. The specialist installer – there are no artisans any more – interprets these as best he can, taking most of his instruction from the label on the product. There is very little knowledge implicit in this dialogue; the end of craftsmanship.

That might seem a bad thing – but it's not necessarily so. As the discussion above indicates, at least in theory, parametric modelling systems enable information of almost any sort about building components to be embedded in the software used to model them. So, provided their authors can express them parametrically, the material properties, handling characteristics and installation methods associated with most building components can be built into their model definitions.

This means that the architect can know all he needs to know about any given building component, at the point when he inserts it into a BIM model. The component 'knows' how it should be handled, how it connects with its neighbours and so on. The architect can focus entirely on the aesthetics of the design challenge. In a sense, the architect becomes the (*archi-tecton*) master builder once again, this time an '*information* master builder'.[2]

The general ability to attach information to building components also means that rules about the behaviour of their real-world counterparts can be embedded in their software definitions. These rules can reflect building codes and regulations, width of corridors, fire resistance of doors and so on. 'Eventually, design handbooks will be delivered in this way, as a set of parametric models and rules.'[3]

The extent to which other industries have been transformed by technologies that enable human knowledge to be embodied in software is discussed in Chapter 8. Construction may have some way to go but, as Eastman and his co-writers put it: 'Parametric modelling transforms modelling from a geometric design tool into a knowledge embedding tool'.[4]

In the AEC area to date, CAD has not generally 'opened up new domains to the architectural imagination'.[5] But it is highly likely that, within the next five years or so, the technology vendors will develop new solutions that go way beyond CAD. Over time, more sophisticated software will become available to support designers for whom the sensual, physical, pencil-on-paper, drawing process forms a large part of their creative method. More advanced drawing tablets, 3D mouse tools, gesture-based interfaces and augmented reality are amongst the technologies that will help designers for whom the process of design is actually carried out in the activity of drawing; 'Thinking with a Pencil'.[6]

Some tools, such as Bentley's Generative Components are already available to help with the design of highly complex, mathematical structures, for example, things like:

> Blobs or metaballs, as isomorphic polysurfaces are sometimes called ... amorphous objects constructed as composite assemblages of mutually-inflecting parametric objects with internal forces of mass and attraction.[7]

Robert Aish's move from Bentley's Generative Components group to work for Autodesk, is intended presumably to extend these capabilities to Autodesk's products, and as noted at 5.3.9, above, to explore further the possibility of bringing about 'a meeting of minds between creative designers who use CAD systems and the software engineers who create these systems'.[8]

---

2 Kolarevic, B., 'Digital Morphogenesis' in *Architecture in the Digital Age: Design and Manufacturing*. Abindon: Taylor & Francis, 2003, p. 27.
3 Eastman *et al.*, p. 41.
4 Ibid., p. 43.
5 Mitchell, W.J., 'Foreword', in Y.E. Kalay (ed.), *Architecture's New Media*, Cambridge, MA: MIT Press, 2004, p. xi.
6 Nelms, H., *Thinking With a Pencil*. Berkeley, CA: Ten Speed Press, 1981.
7 Kolarevic, p.21.
8 Aish, R., 'Extensible Computational Design Tools for Exploratory Architecture', in Kolarevic, p. 245.

The place where human imagination and the computing machine mesh most closely is arguably in the work of Frank Gehry. Jim Glymph, who until recently was technical director of the practice, developed a set of techniques for turning Gehry's fundamentally analogue way of working into digital form. The approach has since been embodied in a product called Digital Project, a specialised architectural version of CATIA, the leading aerospace CAD/CAM system. Being a true CAD/CAM system, with Digital Project 'the design information *is* the construction information', as Kolarevic puts it.[9]

A final point is worth making here about the ability of parametric components to carry useful information. Once a given component is inserted into a model it becomes usable as a hook to which a wide range of other information can be attached. The modelling systems generally keep track of any changes to the component that take place in the course of the design, which is very useful in itself. But this capability becomes particularly useful as a way of keeping track of the component through the construction process and on into its in-use service. For example, as discussed in Chapter 4, the ability to track installation progress at the level of the individual building component will greatly aid in the delivery of projects on time and to budget. The ability to manage the completed building at the component level will enable buildings to be managed through their life-cycles much more economically and much more sustainably.

6.2 Construction project software

There are scores of software packages available for firms in the construction industry. They include general purpose business systems, like accounting and HR applications, as well as more industry-specific products like practice management and billing systems. The focus in this book is on systems for dealing with the technical aspects of construction projects: drawings, specifications, schedules, calculations, reports, RFIs, instructions, scope definition, budget and programme. All of this information, as generated in the course of a building project, is based on and originates in the architectural design. The architecture forms the template from which everything else takes its shape. If the information generated in the architectural design is of poor quality, that will be reflected in all of the other information generated in the project.

The fundamental objective in using BIM is to achieve dramatic improvement in the quality of information created and used on the project. The fundamental requirement of a BIM project is therefore that at least the architectural design is carried out using a true architectural BIM modelling system. Architectural tools are therefore first on our list. However, there are a number of other true, BIM authoring tools – in the sense defined above – available for the other main disciplines: structures and MEP. The list at the time of writing is given in Table 6.1.

It is worth noting that a number of these tools are actually what might be called super-BIM systems. This is because they go beyond basic modelling and can be

---

9 Kolarevic, p. 7.

Table 6.1 BIM authoring applications

Architecture	
Autodesk	Revit Architecture
Bentley	Architecture
Gehry Technologies	Digital Project
Graphisoft	Archicad
Structural engineering	
Autodesk	Revit Structures
Bentley	Structural Modeller
Design Data	SDS/2
Tekla	Tekla Structures
AceCad	StruCad
Mechanical, electrical and plumbing engineering (MEP)	
Autodesk	Revit MEP
Bentley	Building Electrical Systems
	Building Mechanical Systems (CADDUCT, DDS)

used directly to generate the forms of data required by computerised numerically controlled machines used in manufacturing and fabrication. This group includes Digital Project, which is a comprehensive CAD/CAM solution for specialist areas of construction and also Tekla X-Steel and StruCad, which while they are not usually used for front-end design are especially effective for structural detailing and fabrication optimisation.

BIM tools address only a part of the industry's need for technical applications. The construction projects software map in Figure 6.3 shows approximately where BIM systems fit in the overall spectrum of construction software packages. This is not intended to be an exhaustive list; nor is it intended to be definitive in terms of the particular point at which individual products are applied. Its main purpose is to demonstrate the large number of applications with which BIM tools might be expected to share data, thus to highlight the importance of the data exchange standards and protocols discussed below.

There is a problem of missing applications. BIM authoring tools can all easily generate schedules of components, arranged in a variety of ways, according to all known standard classification systems. These take-offs can easily be passed to contractors and suppliers for pricing. However, it's important to recognise that the physical components of the building are the contractors' deliverables; their outputs. In order to calculate his price for a given component a contractor must estimate the cost of his inputs: plant, labour, materials and overages for example. Innovaya Composer and Vico are two products which can be used for that purpose in the USA; no such products currently exist in the UK. This is one example of the missing apps problem. Solutions will no doubt appear in due course, but meanwhile contractors and others are having to develop workarounds and fixes to enable their conventional systems to take advantage of BIM-generated data.

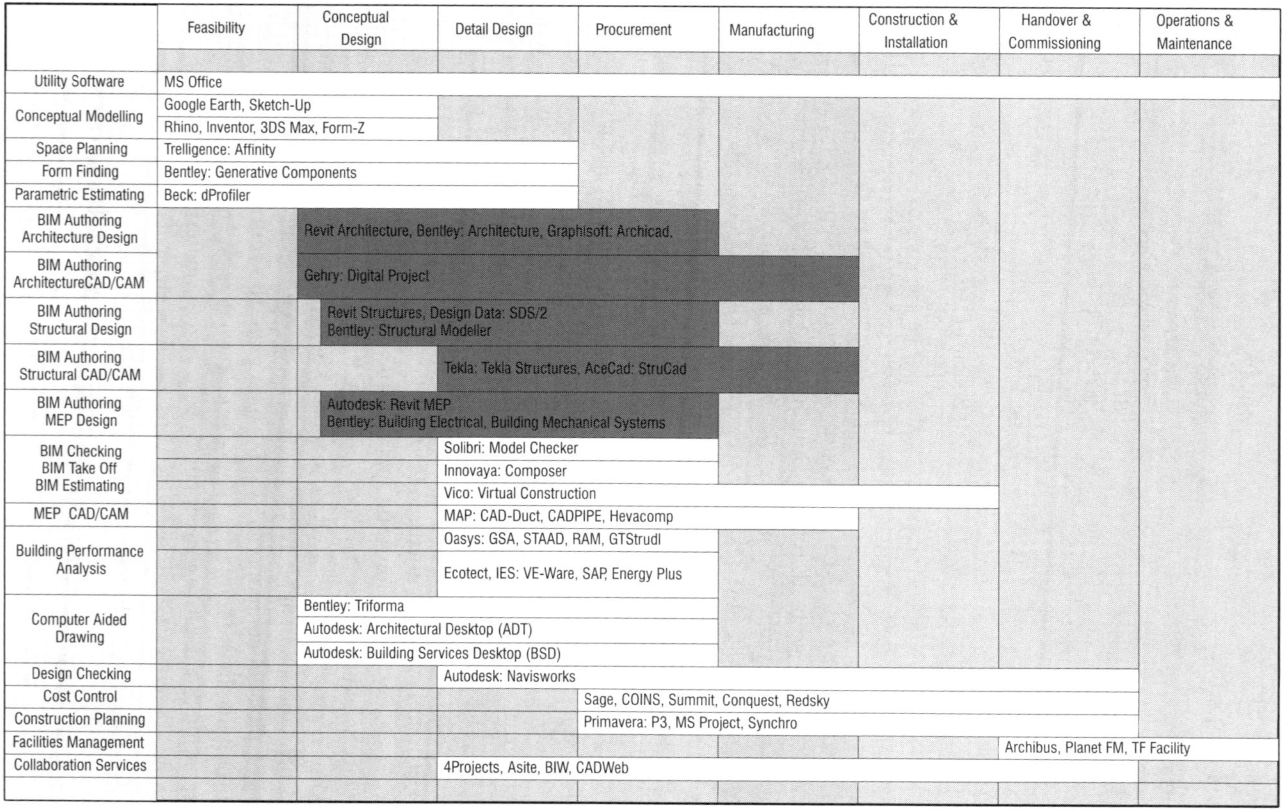

	Feasibility	Conceptual Design	Detail Design	Procurement	Manufacturing	Construction & Installation	Handover & Commissioning	Operations & Maintenance
Utility Software	MS Office							
Conceptual Modelling	Google Earth, Sketch-Up							
	Rhino, Inventor, 3DS Max, Form-Z							
Space Planning	Trelligence: Affinity							
Form Finding	Bentley: Generative Components							
Parametric Estimating	Beck: dProfiler							
BIM Authoring Architecture Design		Revit Architecture, Bentley Architecture, Graphisoft: Archicad.						
BIM Authoring ArchitectureCAD/CAM		Gehry: Digital Project						
BIM Authoring Structural Design		Revit Structures, Design Data: SDS/2						
		Bentley: Structural Modelier						
BIM Authoring Structural CAD/CAM			Tekla: Tekla Structures, AceCad: StruCad					
BIM Authoring MEP Design		Autodesk: Revit MEP						
		Bentley: Building Electrical, Building Mechanical Systems						
BIM Checking			Solibri: Model Checker					
BIM Take Off			Innovaya: Composer					
BIM Estimating			Vico: Virtual Construction					
MEP CAD/CAM			MAP: CAD-Duct, CADPIPE, Hevacomp					
			Oasys: GSA, STAAD, RAM, GTStrudl					
Building Performance Analysis			Ecotect, IES: VE-Ware, SAP, Energy Plus					
Computer Aided Drawing		Bentley: Triforma						
		Autodesk: Architectural Desktop (ADT)						
		Autodesk: Building Services Desktop (BSD)						
Design Checking			Autodesk: Navisworks					
Cost Control				Sage, COINS, Summit, Conquest, Redsky				
Construction Planning				Primavera: P3, MS Project, Synchro				
Facilities Management							Archibus, Planet FM, TF Facility	
Collaboration Services			4Projects, Asite, BIW, CADWeb					

Figure 6.3 Construction project software map

6.3 Information management on BIM projects

Information management on BIM projects is not a great deal different from good project information management practice generally. Successful implementation of the Building Information Modelling approach on a project requires careful attention to three important issues:

- the structure of the project organisation and the type of procurement strategy;
- the implementation of agreed exchange file formats for all key applications, identifying the types of information that might be exchanged between the different applications;
- the implementation of agreed information interchange protocols, identifying the originator of each type of information and the status or level of detail it should contain at each interchange point throughout the duration of the project.

There are many sources of advice and information about each of these issues. The references cited here are almost all comprehensible and useful to the non-specialist. However, it is very easy for the general manager to become overwhelmed by the technical material they contain. It is important to recognise that the various guides, standards, specifications and so on are usually written in such a way as to cover all potential situations and circumstances. They should be used selectively and with care, to ensure that a level of good order is achieved that is commensurate with the size and complexity of the project at hand. The information management strategy for a project should be as simple as possible, but no simpler, as Einstein might have said.

6.3.1 BIM implementation strategy

In the implementation scenario envisaged here, the architect creates the baseline architectural model of the building and publishes it to the members of the team. (Ideally, even at the very earliest stages of design, the team should include the main contractor or construction manager.) Each of the other design firms uses this as the basis on which to create its own discipline-specific sub-model. Periodically these sub-models are brought together to form a central reference model where the different contributions can be checked for gaps, clashes and other anomalies. At least initially, this process is envisaged as taking place in round-table, face-to-face meetings with all relevant members of the team present. (Obviously, this can also be done using video-conferencing and application-sharing technologies, but it makes sense for at least the initial series of meetings to be held in person.)

This is how effective design coordination meetings are conducted today, in a world of paper drawings. That is very much the point. BIM technology is not yet at a stage of maturity and capability where it would make sense to recommend the deployment of a single shared model, supporting fully real-time concurrent design by a multiplicity of firms, at a variety of locations. The merits of this mode of operation in AEC design are unproven, and the person-to-person interaction involved in the traditional way of working is in itself desirable.

The key aspect of BIM implementation is that it be carried out within a well-documented, overall information management strategy. The highest level BIM strategy document should be no more than a single page in length, should contain as little technical material as possible, but should be precise in expressing the strategic objectives and commitments required to implement BIM effectively on the project. As with most other elements of a project execution plan, the BIM strategy must be a 'cascade' document – with appropriate counterparts at the different levels of the overall project organisation.

6.3.2 Project organisation and procurement strategy

There are a number of ways in which the overall project can be organised to take advantage of the Building Information Modelling approach. At present, and broadly speaking, these all require that a basic level of collaboration be achieved at least amongst the consulting firms involved, but preferably also including the main contractor. The contractor should be familiar with the model(s) and should attend the coordination meetings, at least in order to provide buildability and specialist contractor input.

The crucial requirement is that, from the beginning, the principals and project managers from each of the key organisations – client, architect, principal consultants and main contractor – should agree explicitly to make information management a strategic issue on the project.

The mainstream UK construction industry has been exploring collaborative approaches to major projects for over 15 years. The first project-wide information management extranet is thought to have been deployed by Bovis Construction, on the Bluewater retail centre project in Kent, on which work started in mid-1996. The extranet concept has expanded greatly since then, with several dedicated application service providers now providing collaboration, document management and other communications services to construction projects.

The success of a BIM implementation on a given project depends on the same sorts of arrangements being in place as are required to implement these services well. That's not to say that if a project is properly set up for collaboration it will also, by default, be able to support a BIM implementation – but it's a good starting point.

Main contractor

One of the most collaborative approaches to project organisation has been in use in North America for about five years, in a form of a procurement philosophy called integrated project delivery (IPD). In IPD the designers, the main contractor and the key sub-contractors enter into an agreement with each other to deliver the project as an integrated virtual organisation. The American Institute of Architects (AIA) provides a comprehensive guide to IPD, including proposed contract forms.[10] The relationship between the partners can be governed by relational contracts of various types, partnering arrangements, project and strategic alliances and specifically created single purpose entities (SPE) agreements. The overall objective is to achieve a sharing of goals and close collaboration amongst the main project team members. Key project targets, including a budget estimate, are agreed amongst the team. The team members are fee-reimbursed for their personnel and other costs, and nominal profit levels are agreed at the outset. But the principal form of reward is a profit share between the client and the team, based on the achievement of agreed project targets and milestones. This usually takes the form of a pain/gain arrangement (over-runs incur penalties) which helps keep team members focused on joint benefits and shared problem solving. Crucially, IPD attempts to eliminate zero-sum gaming amongst the team members. IPD lends itself well to the agreement of data exchange standards and protocols as discussed below.

Similar approaches have been pursued in the UK and elsewhere for many years. Project alliances have been used for many years in the UK North Sea oilfield construction sector and more recently elsewhere, for example the Australian state of Victoria.[11] Many major organisations elsewhere in UK construction have participated to some extent in project (and strategic) partnering, as promoted by, amongst others, the Construction Industry Council (CIC).[12] So, the UK construction industry is familiar with these forms of project organisation, though not with the use of BIM in carrying them out. As noted in Morledge *et al.*, however, experience of partnering, alliances, frameworks and suchlike has not been universally favourable in the UK,[13] so some work may remain to be done to achieve the optimal combination of BIM methods with this more collaborative approach to contractual relations. These issues notwithstanding, a partnering-based project can be expected to cope relatively easily with the negotiation of BIM standards and protocols.

The form of project organisation in which BIM has been most widely used is probably design and build (D&B). The key feature of D&B is the transfer of almost all project delivery risk from the client to the D&B contractor. As noted in Section 2.1, the latest RICS / Davis Langdon 'Contracts in Use' survey shows that between 1985 and 2007, the proportion of projects delivered under D&B forms of organisation increased from 8 per cent to 32.6 per cent.[14] Whether this rate of growth has continued through the current recession remains to be seen,

---

10 American Institute of Architects, *Integrated Project Delivery: A Guide*, Version 1, 2007, http://www.aia.org/contractdocs/AIAS077630 (retrieved 1 June 2010).
11 Department of Treasury and Finance, *Project Alliancing: A Practitioner's Guide*, Melbourne: Department of Treasury and Finance, State of Victoria, 2006.
12 Construction Industry Council, *A Guide to Project Team Partnering*, 2nd edition, April 2002 available from http://www.cic.org.uk/activities/partnering.shtml (retrieved 4 December 2009).
13 Morledge, R., Smith, A. and Kashiwagi, D.T., *Building Procurement*. Oxford: Blackwell, 2006, pp.96–7.
14 RICS, Davis Langdon, *Contracts in Use: A Survey of Building Contracts in Use During 2007*. London: RICS, 2009.

but it would seem to be likely that D&B will continue to be used on a significant proportion of UK projects. Because in this arrangement the consulting team are all employed directly by the main contractor, he can influence them strongly in the systems they use and the standards and protocols deployed on the project. D&B will probably be the most important proving ground for BIM working in the coming years.

One of the most favourable environments in which to deploy BIM methods is in management forms of contract: construction management, where the client contracts with and pays the specialist contractors directly; and management contracting, where the management contractor holds the contracts and makes the payments. Management forms of contract were introduced initially to enable fast-track operation in which individual construction contract packages can be procured as soon as the design of the relevant packages is complete, rather than having to wait for the entire building design to be completed, as is necessary when the main contract is awarded on a traditional lump sum basis. In both of these types of arrangement, the main contractor, referred to from here on as the construction manager (CM), becomes part of the professional team, paid on a cost plus fee basis, and engaging collaboratively with the other members of the team.

The CM is usually appointed early in the project, typically during the conceptual or early detail design stages. His main role is to plan and monitor the design and construction stages in detail, to test the design for buildability and to procure and manage the specialist contractors. The essential contribution of the CM is detailed knowledge of the design and fabrication techniques and capabilities of the key specialist contractors, particularly the structures, cladding and M&E contractors. This is necessary in particular to ensure that the flow of information from these processes back into consultants' design programme, is planned and managed effectively.

Projects on which the main contract is awarded on a traditional lump sum basis are not considered appropriate for the application of BIM techniques at the present time. This is mainly because of the adversarial nature of the relationship between the main contractor and client to which such contracts tend to give rise.

So, in the UK at least, the use of BIM methods is likely to be limited to projects using partnering, design and build and management forms of contract for the main contractor. Note however that, at 58.8 per cent by value of all contracts, this represented a substantial proportion of the total output of the industry in 2007, as the RICS study of contracts in use shows.[15]

Specialist contractors (term used here includes sub-contractors, trade contractors etc.)

It may seem perverse to suggest that although the main contractor should be appointed on a non-confrontational/collaborative basis, the specialist contractors should be selected on a competitive, fixed price, lump sum basis. The strategic

---

15 Ibid., p. 7. Chart 2: Partnering, 15.6 per cent; Design and Build, 32.6 per cent; and Management Forms, 10.6 per cent.

arguments in favour of the use of competitive tendering of the actual construction work, as opposed to the largely administrative work of the main contractor, were laid out in Sections 3.3 and 4.3 above.

In an ideal world, using a BIM model, the entire building would be fully designed and tested before going to the market for construction contracts. The design and scope information provided in tender packages would therefore be complete and unambiguous, and would be readily verifiable as such. The bidders' responses would be clear and comparable and the winner's tender would be guaranteed to be the lowest price – for the specified product or service. In other words, the client would get the best available value for money from a competent contractor or supplier. Assuming that the consulting firms are contracted on a full services, cost-plus basis, it should be possible in this scenario to procure, price-competitively, more or less all of the actual construction work: 80–85 per cent of the total cost of most projects.

Unfortunately, the world at the moment is less than ideal in this respect. Figure 6.4 illustrates the main problem. Certain of the specialist contractors and suppliers are thought to possess key construction or product information that the design team cannot be expected to know about in advance. These companies must therefore be appointed, uncompetitively, at an early stage, so as to enable their particular information to be incorporated into the overall design. Given that cladding and mechanical services, the main trades to which this applies, can together comprise over 60 per cent of total construction value, this problem can cut across the goal of competitive tendering pretty seriously.

There are two possible solutions. First, it may be feasible to appoint the consultants on a true, full services basis, avoiding recourse to 'design intent' expediencies. For example, MEP consultants might be contracted to provide fully dimensioned and coordinated, detailed mechanical services design, as part of their contract, without reference to particular trade contractors or suppliers. (This approach might actually be expected to result in more professional, less contractor-biased solutions.) Alternatively, elements like external cladding, for example, might be procured on a two-stage basis. In this arrangement an initial cost-plus contract is placed for the development of a detailed non-proprietary design. The construction of this design is subsequently issued for competitive tender in the open market.

These problems will both go away as true BIM design tools become pervasive. First, as the range of parametric BIM components becomes more complete, and as individual components become richer in content, designers will be able simply to insert a given component into the building model and know that all its attendant construction detailing will follow. The resulting model will generate fully detailed designs and construction scopes of work, with no requirement for contractor design input.

Also, in order to be competitive in this new marketplace, equipment and component vendors will have no choice but to make the details of their products available as intelligent models. This may be through the use of online services like Autodesk Seek, the NBS, or other services that keep secure the intellectual

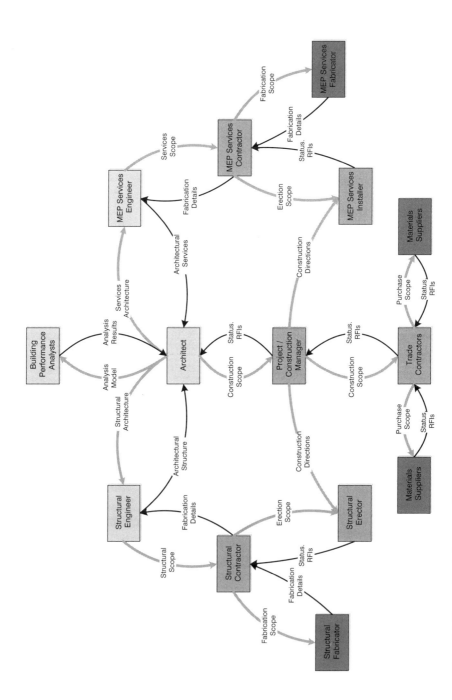

Figure 6.4 Key project information flows

Table 6.2 Content of key project information flows

Flow title	Flow content
Structural architecture	The architectural information required by the structural engineer, in order to carry out the structural analysis and design
Structural services	The architectural information required by the services engineer, in order to carry out the services analysis and design
Architectural structure	The information generated by the structural engineer that is required for the architectural design
Architectural services	The architectural information generated by the services engineer that is required for the architectural design
Analysis model	The architectural information required by the various building performance analysts
Analysis results	The results from the analyses required for the architectural design
Construction scope	Information required to specify contractors' scope of work
Status/RFIs	Status reports and requests for information, from contractors upwards
Structural scope	Information required to specify structural contractor's scope of work
Services scope	Information required to specify structural contractor's scope of work
Fabrication scope	Information required to specify fabricator's scope of work
Fabrication details	Information generated by fabricator that impacts architectural design
Construction directions	Information governing the operations of contractors on site
Purchase scope	Information required to specify scope of bought items

property embedded in vendors' models. If they don't, specifiers will not learn of their existence or will simply ignore them.

The diagram in Figure 6.4 illustrates some of the challenges involved in attempting to achieve efficient and logical flows of technical information on large projects. Table 6.2 provides an outline of the content of each of these information flows.

6.3.3 Data exchange standards – interoperability

Interoperability has become a bit of a bugbear in the BIM world. It has never been dealt with satisfactorily in relation to conventional CAD applications. There are two sets of problems with interoperability. First, making clear who is responsible for producing what design information to what level of detail at what point in the development of the design. In other words, what design information can I expect you to produce, at what level of detail at any particular point in time? That is a problem of business process interoperability; do our two firms understand exactly how each other does business? It is essentially a problem of design management and coordination.

Then there is the problem of interoperability as a data-processing problem: how can I pass a particular piece of information from my system to your system and ensure that your system 'understands' that information to mean exactly the same thing as my system does? In trivial terms, if I pass something like '2+2=4' from my machine to your machine, will your machine be guaranteed always to understand the statement as '2+2=4' and not something like '2+2=5'? So two steps are involved: ensuring the communicating systems code statements in the same way, and ensuring that they both act on the data in exactly the same way. A good, short, non-technical discussion of the issues relating to interoperability and open standards is provided on Wikipedia.[16]

A strong distinction is made here between the low-level technical issues involved in data exchange between different computer applications, and the more complex social challenges involved in the communication of technical information between different organisations. The first is covered here under the heading of data exchange standards; the second is addressed in the following section on information interchange protocols.

The efficiency of most computer applications is heavily influenced by the way in which they internally store and organise the data they work with. Particular types of data manipulation processes, particular types of calculation and of graphical presentation, require that the data they use be stored in particular formats and in particular structures, in order to maximise their speed of response and to optimise their requirements for computing resources. The developers of applications therefore design their systems' data formats and structures to reflect this. The result is that no two applications store and manage their data in the same way.

As anyone who has worked with even the most basic of data systems will recall, trying to export information, e.g. from one simple Excel spreadsheet to another, can be ridiculously difficult to do. The problem is usually that the data fields (the columns of data) are called different things, or the format of the data (whether it's a numerical field or a text field, for example) is different between the two spreadsheets. This is a very simple example. It reflects two problems: how data entities are identified, and the type or nature of the entities. In this case, both spreadsheets were created in Excel. Another more profound problem arises when the target and destination systems are not the same application. Excel stores its information in relatively simple sets of tables – databases of rows and columns, corresponding in structure to the shape created by the user. However, other applications use different, sometimes far more complex structures for storing data.

BIM modelling systems and the other applications in the project software map in Figure 6.3 are all specialised and highly optimised in terms of the data they

---

16 http://en.wikipedia.org/wiki/Interoperability (retrieved 1 November 2010).

work with and the storage methods they use. This creates a serious problem for project teams wishing to exchange and share information. It was recognised as such very early in the development of computer-aided design. Weisberg records how a colleague of Steve Coons at MIT, in 1960, wrote about the need for systems to be able to share a wide variety of types of information in order to provide a comprehensive CAD capability.[17]

This area of computing is a bit of a swamp; but there are three basic ways in which CAD systems can exchange information:

- Through the use of one-to-one translators, in which each system in a pair of sending and receiving systems deploys a translator program so as to be able to read and write files in each other's format. This means that every system in a particular domain needs to have available a translator for every other system; so the approach is relatively rarely used.
- Through the use of a proprietary file format, in which one system vendor publishes a file structure for which other applications vendors can develop routines that enable them to read and write the first vendor's files. Autodesk's DXF is the most widely used CAD file format of this type.
- Through the use of a neutral exchange format for a particular domain, in which each system in the domain is able to read and write data in the neutral format. Each system thus needs to be able to read and write files only in the neutral format.

Other techniques exist. Companies such as Autodesk publish applications programming interfaces (APIs), which enable programmers of specialist analysis packages, for example, to write applications that can access directly the data contained in AutoCad files. This is not quite the same as generalised data exchange, which depends on one of the three methods outlined above.

Neutral exchange file formats

The first systematic attempt to deal with the problem of data exchange between different CAD systems was in 1979 when the US Department of Defense and the US National Bureau of Standards, now the National Institute for Standards and Technology (NIST) contracted with Boeing, General Electric and a number of CAD companies to developed a neutral file format which was called IGES (Initial Graphic Exchange Specification). IGES has passed through a number of revisions, reflecting advances in CAD technologies. The last published revision was Version 5.3 (1996). IGES remains one of the most widely used data exchange formats, particularly in the mechanical CAD arena.

In the mid-1980s, supported by the US military Computer-Aided Acquisition and Logistics Support (CALS) program, seeking ways of tracking military equipment

---

17 Weisberg, D.E., *The Engineering Design Revolution: The People, Companies and Computer Systems that Changed Forever the Practice of Engineering.* http://www.cadhistory.net/ (retrieved: 24 October 2010), Section 3, p. 11.

and related spares and consumables, work started on the development of PDES (Product Data Exchange Specification). As the name suggests, PDES addressed more than just the design information about an object. It was intended that PDES should make it possible capture and exchange more or less every piece of information that might be generated in relation to any given product throughout its design, manufacture and expected life in use. This hugely ambitious initiative was subsequently merged into an even more ambitious project run by the International Standards Organisation (ISO), and called Standard for the Exchange of Product (STEP) data. STEP is a stunningly complex enterprise that seeks to enable every conceivably useful attribute of every significant product in the modern economy to be captured and shared by pretty well any imaginable computing device.

The first published STEP specification of near-relevance to construction, ISO 15926: 'Process Plants including Oil and Gas facilities – Life-Cycle Data', after nearly 20 years work, was finalised in 2007.[18]

Of more immediate relevance to building construction has been a program called the International Alliance for Interoperability (IAI), which was set up, largely at the instigation of Autodesk, in 1995. IAI, recently re-named BuildingSMART, has produced a series of standards called Industry Foundation Classes, whose latest revision is at version IFC2.3 (2006), with a new version IFC2.4 released in preliminary form in May 2010. The IFC work has been endorsed by ISO as a Publicly Available Specification, ISO/PAS 16739; its current status is 90.92 – 'International Standard to be Revised'. All of the BIM authoring tools and many other specialist construction applications support at least IFC2.3. So in theory, it should be relatively straightforward to exchange data between these systems.

An important STEP-based standard for the structural steel industry, CIS/2, was developed by Andrew Crowley and Alastair Watson of Leeds University. CIS/2 (CIMsteel Integration Standard Version 2)[19] was the main deliverable from a major EU-funded R&D project called CIMsteel which was completed in 1998. CIS/2 is an exemplary data exchange standard that satisfies the requirements of the entire structural steel industry, from design and analysis through fabrication and installation. CIS/2 was adopted by the American Institute of Steel Construction (AISC) in 1998, and most of the relevant software vendors are now CIS/2 compliant.

(It is instructive to observe how the CIS/2 standard came about: realistic ambitions, acute focus, and active involvement of its user community, leading to a fully workable standard in a reasonable period of time, with comparatively little effort. Though based closely on ISO methods and technologies, and although a draft ISO standard had been initiated (ISO 10303 – AP 230) it has not been considered necessary for CIS/2 to be promoted as an ISO standard and AP 230 has been abandoned. Some issues relating to the IFC standards are discussed in Lipman's NIST paper comparing the capabilities of the IFC and CIS/2 data models.[20])

---

18 http://en.wikipedia.org/wiki/ISO_15926 (retrieved 4 May 2010).
19 Crowley A. and Watson A. *CIMsteel Integration Standards*, Release 2, SCI-P-268. Ascot: The Steel Construction Institute, 2000.
20 Lipman, Robert R., 'Details of the Mapping between the CIS/2 and IFC Product Data Models for Structural Steel', ITcon 14 (2009): 1–13. http://www.itcon.org/2009/01

In the case of MEP systems, significant deficiencies persist both in the range of available BIM components, and in the relevant data exchange capabilities. The sector is very fragmented, but a focused CIS/2 type exercise, perhaps based on the experience of developing ISO 15926, would seem to be a useful project.

It is important to bear in mind that data exchange is not just a capability required for today's purposes and applications. Accurate, practical data exchange methods will also ensure that material which is archived today will be restorable and reusable 10 or 20 years from now.

The real world – proprietary formats

In the present circumstances, it is highly unlikely that any private-sector organisation – construction firm, client or software vendor – can afford to wait the 20 or so years it takes to achieve a fully ratified international standard in this area. The IFCs are making good progress and, together with CIS/2, will satisfy the requirements of many organisations, including those in the public sector. However, the fact is that a file exchange between any two systems, particularly one that involves translations both ways via a third, neutral format, is almost inevitably going to be subject to some loss of information and some degree of misinterpretation. An interesting solution to these problems is proposed by Rappoport.[21]

Arguably the most expeditious solution is surely to eliminate the need for the third format. This is what has happened in the area of conventional AEC CAD, with Autodesk's DXF format; problematic for the public sector and competition purists perhaps, but greatly to the advantage of the construction industry at large. The announcement in 2008 by Autodesk and Bentley that they will in future share each other's formats is an important move in this direction for the BIM community. It would seem reasonable to suggest that, should Autodesk be seen to abuse its position in this regard, competition regulations might be invoked by its competitors – as happens regularly with Microsoft, for example.

In the short term, a pragmatic approach will be taken to the issue of interoperability. The people participating in BIM design teams will be well experienced in the exchange of traditional CAD files. For the most part, similar operating procedures will apply with BIM designs.

6.3.4 Information interchange protocols

The BIM approach to the management of project information depends heavily on the commitment of the main firms to ensuring that information flows on the project should be as efficient and as responsive as possible. The priority is to get people who create information to think about how other people will want to

---

(retrieved 4 May 2010).
21 Rappoport, A., 'An Architecture for Universal CAD Data Exchange', *Proceedings Solid Modeling '03*, June 2003, Seattle, WA: ACM Press.

use that information and, therefore, how they should prepare and present their material for that purpose.

This requires that the project team agrees to share information in agreed formats and according to an agreed set of protocols. The overall needs of the project are agreed to take priority over the internal standards of individual firms. To achieve this level of cooperation is not easy; but neither is it prohibitively difficult. The five key guidelines are:

- Get commitment from the principles and project managers of the main firms, at the very beginning of the project, preferably during the conceptual design phase.
- Be realistic. In particular, resist the impulse to over-automate information exchanges. Selective manual interventions can often lead to more robust processes than elaborate automatic file processing routines.
- Everybody does not need to be able to edit all of everybody else's information all the time. In fact, the vast majority of people on a construction project only ever need to be able to read other peoples' information. So again, be selective.
- Any firm that incorporates information created by any other firm in its own design, and publishes the result, thereby takes full responsibility for the accuracy and truth of the published information.
- The information standards developed for the project need to be at least as effective as those of the individual participating firms.

The information management challenge starts with understanding the actual flows of technical information around the project: who provides what information, to whom, and when? In that sense the flows encountered on a BIM project will be essentially similar to those on any well-organised, conventional, collaborative project – somewhat similar to the arrangement illustrated in Figure 6.4. This diagram is intended to represent the point-to-point exchanges of technical information; it does not reflect the contractual links or administrative overhead associated with these transactions.

The basic workflow amongst the designers is for the architect to issue the 'structural' architecture to the structural engineer and 'services' architecture to the MEP consultants. This is the material that those firms identify as being necessary to carry out their work. The consultants carry out their analysis, calculations and design and return to the architect their proposed changes to his material. This process is repeated as required.

The workflow amongst the contractors requires the construction manager (CM), or main contractor as the case may be, to issue scope information and construction directions to the specialist erection and installation contractors. They feed back the status of their work at regular intervals and also raise queries or requests for information with the CM. Contractors who procure, or carry out themselves, the fabrication of components or assemblies of components, are also required to feed back relevant fabrication details to the appropriate design firms.

These are obviously highly simplified versions of the reality of construction communications. In particular, rather than being direct one-to-one exchanges as suggested here, the flows of information between consultants and the corresponding fabricators will usually have to be routed via the construction manager. This routing is usually necessary for contractual and administrative reasons. The underlying technical information exchange is as shown in the diagram – the construction manager will only rarely need to intervene in the exchange at that level. Figure 6.4, or a similar project-specific map, will identify the information links that exist between the various firms on the project. The next step is to identify the particular types of information that will pass along each of the links at particular points in time during the project.

As with conventional collaborative projects, once the overall scheme is defined, it can generally be advantageous to complete the detail design on a construction trade-by-trade basis. The feature that is particularly useful about this approach is that it focuses design production very clearly onto the outputs, or deliverables, required at each key stage in the overall project process. The RIBA Plan of Work[22] provides one way of identifying these stages. Figure 6.5 shows the key deliverables for each discipline, for each Plan of Work stage on a typical building construction project.

A more general deliverables-based breakdown is given in Table 6.3. This sort of breakdown focuses on the immediate use to which deliverables created in each of five project stages will be put. The approach helps to clarify the network of relationships between design activities. It is simple but robust and has been used in parts of the process plant design industry for some time. Intermediate stages can be introduced here if required. The level of detail of the design at each stage may vary from trade package to trade package. It should not be the originator of the information who gives it its status; this should be the responsibility of the design manager for stages A and B and the construction manager for subsequent stages.

6.3.5 Contracts, model ownership, insurance and intellectual property

BIM is an innovative approach to construction, which requires a degree of collaborative intent on the part of the client, consultants and main contractor to work successfully. So this approach is unlikely to be deployed on projects where the main construction contracts are awarded on a competitively tendered, fixed price, lump sum basis, for some time. (But that time will come – see Chapter 9.)

In the context of any of the generally collaborative forms of contract considered above, BIM does not change any of the contract, insurance or IP issues fundamentally. The general principle is that BIM just creates information. It is in electronic format, but this is not in any fundamental legal, contractual or professional sense different to the information generated in a paper world. No

---

22 For 'An Outline of the Plan of Work' see http://www.pedr.co.uk/textpage.asp?menu=1a &sortorder=130&area=main (retrieved 4 May 2010).

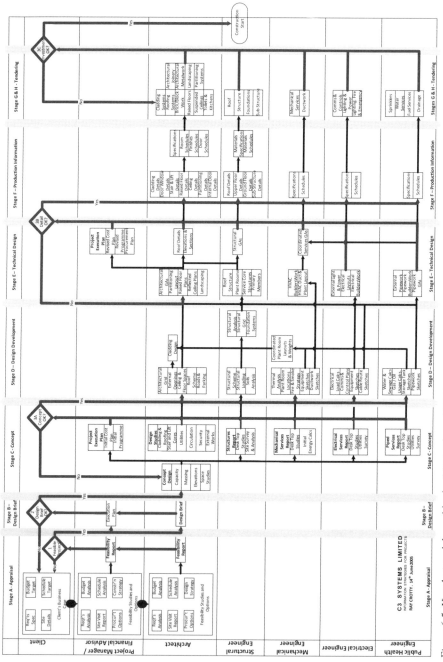

Figure 6.5 Key design deliverables by RIBA Plan of Work stage

Table 6.3 Design stages: indicative level of detail

Stage	Level of detail
Stage A	Preliminary design work: information created for internal use only
Stage B	Issue for Design: design material released by one designer for use in the design work of another department or firm
Stage C	Issue for Tender: design documentation released for the purpose of tendering a specific trade package
Stage D	Issue for Contract: documentation released to provide the contractual scope of work of a specific trade package
Stage E	As Built: documentation, including survey information, recording the as built condition of the relevant building element or elements

special forms of contract are required, though some have been developed in the USA, as noted above.

Similarly with the issue of 'ownership' of the model. Each firm 'owns' its particular discipline-specific part of the overall model. Each firm is responsible for the accuracy and completeness of any information it publishes, whether it be from its model or any other of its sources of information – just as in the paper world. In the short term, on traditional contracts, it is unlikely that contractors will accept models alone as contract documentation; drawings will have to be generated from the various models for procurement and construction issue purposes. However, on the basis of the interest currently being shown by the major players, it is likely that most contractors will accept copies of coordinated reference models for 'information' purposes. In a collaborative environment, of course, this issue does not arise; the main contractor will take ownership of the finalised coordinated reference model(s) and will be expected to base the construction documentation on the content of those model(s).

Any professional firm intending to work on a BIM project *must* inform its professional indemnity insurer before starting work on the project. But there is no indication that insurers are antagonistic to the use of BIM by their customers. On the contrary, it is likely that insurers will, relatively soon, start to offer reduced premiums to firms using BIM

As with other aspects of the 'BIM versus paper' debate, intellectual property (IP) issues tend to become somewhat inflated. In reality, as with the other issues considered here, nothing much changes. The client buys a one-time licence to use the designers' ideas, components and models for the purpose of building the particular building in question. No one else may take ownership of or use any of the material provided by an individual firm without the firm's permission. Digital watermarking techniques will be available to help guard against this problem.

There are currently two tricky issues in the IP area. First, because the BIM vendors do not – probably could not – provide fully comprehensive sets of component families 'out of the box', many design firms are creating their own. These 'home-made' families may not be as well designed as their 'shop-bought' equivalents, in that they may lack the structure, attributes and other characteristics that the vendors apply to their products. This may exacerbate interoperability problems during the course of the project.

The second problem derives from the fact that the home-made families are created using vendors' tools and high-level scripting or programming languages. Staff in engineering and design firms are expert engineers and designers, not computer programmers. Because these parametric components are quite complex and have many possible uses, a component that seems to behave correctly on its designer's screen, that seems to print out properly on a drawing, may not actually perform correctly, in an analysis application for example. Firms need to be very careful that their people are applying the highest possible standards, both of programming and of engineering, in undertaking this work.

6.4 Sources of BIM implementation guidance

Despite the fact that the BIM approach has been applied to relatively few projects, only relatively recently, a significant amount of useful material is already available to guide project teams interested in going down the BIM route.

Perhaps the first thing to grasp about the whole BIM phenomenon is that BIM is just one stage in a continuum, progressing from the most basic design techniques through to a fully integrated design, manufacture, construct world where construction becomes part of the manufacturing sector. Bew and Richards attempted to encapsulate this evolutionary process in their BIM ramp diagram of 2008 (Figure 6.6). Their assessment that 95 per cent of UK users sit in Phase 0 using 2D drawings is probably still valid. But evidence is growing that the industry is slowly moving up the ramp, as Chapter 7 will show.

Much of the available technical documentation on BIM implementation is based on conventional CAD standards, which, given that the technical aspects of operating a BIM project will generally be the responsibility of CAD managers, is as it should be. The *BIM Handbook*[23] is the bible in this context. Its breadth of coverage, readability, respect for standards and overall common-sense approach is highly commendable. As with the other American sources referred to below, the reader must bear in mind its North American focus. For example, the American practice of ending the consultants' services at Scheme Design – approximately what the RIBA Plan of Work calls Stage D, Technical Design – and having the specialist contractors effectively provide everything from that point onwards. British practice has been moving towards that approach in recent years, but it is still not common for the contractors to provide much more than construction/shop drawings and method statements.

Perhaps the first technical port of call for UK readers is the Construction Project Information Committee (CPIC),[24] including the PIX Project protocol documents. Also consider the Avanti project deliverables at Construction

---

23 Eastman *et al*.
24 http://www.cpic.org.uk/en/current-projects/bim/building-information-modelling.cfm (retrieved 4 May 2010).

Building Information Modelling 107

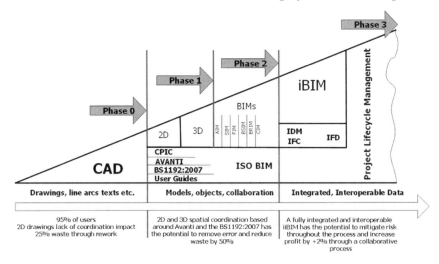

Figure 6.6 Bew–Richards BIM diagram

Excellence.[25] The recent British Standard, BS1192:2007, available from the BSI,[26] provides useful technical advice on processes and naming conventions, as does the AEC(UK) group's somewhat more accessible *BIM Guide*.[27] To date the most exhaustive of BIM standards documents is the US National Institute of Building Sciences' 'National BIM Standards'[28] – that is, the NIBS' 'NBIMS'. Arguably the most useful management guide to actually doing BIM on a project is provided by the American Institute of Architects' *BIM Protocol Guide*.[29] And one of the best current sources of all sorts of BIM-related information is an online journal called *AECBytes*[30] – highly recommended.

But, to repeat the advice offered at the beginning of Section 6.3, keep it simple. Learn from these sources, but apply only that which is relevant and necessary for the project at hand.

6.5 Conclusion

The general power of BIM authoring tools to create very high quality, computable information was outlined above. This quality improvement is not a gradual, incremental improvement along a continuous progression; it represents a complete and profound break with what has gone before. The actual nature of

---

25 http://www.constructingexcellence.org.uk/ceavanti/default.jsp (retrieved 4 May 2010).
26 http://www.bsi-global.com/en/Shop/Publication-Detail/?pid=000000000030163398 (retrieved 4 May 2010).
27 http://aecuk.wordpress.com/ (retrieved 4 May 2010).
28 http://www.wbdg.org/bim/nbims.php (retrieved 4 May 2010).
29 AIA, *Building Information Modeling Protocol Exhibit Document E202™*. Washington, DC: AIA, 2008.
30 http://www.aecbytes.com/ (retrieved 4 May 2010).

the information has changed. When design information of this quality was first introduced into manufacturing 40 years or so ago, it transformed industries in that sector. When retailers started to gather precise information on stock levels and movements, they brought about a revolution in the high street.[31] Both of these examples of sudden, discontinuous change in the evolution of industries, driven by dramatic improvement in information quality are discussed in the next chapter.

The key point is that, in both of these cases, the beneficial effect depended on improved information quality *and* improved information sharing between business functions within firms and between individual firms in the value chain. The technology gave rise to the improved information quality, but the improved information sharing took significant and sustained management intervention.

---

31 Brown, S.A., *Revolution at the Checkout Counter.* Cambridge, MA: Harvard University Press, 1997.

7 BIM – the current state of play

7.0 Introduction

Following on from the introduction to BIM in Chapter 6, the purpose of the present chapter is to map out current BIM usage patterns, mainly in the UK, but also including reference to France, Germany and the USA. Two principal sources of information are used: recently published surveys and a number of case studies. The surveys provide an idea of the level of uptake amongst the different industry players in each of the geographical areas. They also indicate how current users view BIM and the benefits they are, or claim to be, deriving from its use. However, for the purposes of this book, it is more important to explore the attitudes of the individual users and their firms, the methods being used on projects and the sorts of effects that are being experienced by the users. So the larger part of this chapter is given over to a series of short case studies that explore those aspects of real BIM today.

7.1 Surveys

The writers of the survey reports, though generally fair and honest in their analysis, tend to use language that praises the more progressive users of BIM technologies (usually architects) and simultaneously castigates the laggards (usually contractors). Similarly, BIM adoption rates in the USA are usually praised as being ahead of those in Europe. This is an issue of language alone. There is no sense in which rates or levels of adoption of BIM should be regarded as being either laudable or culpable. And there is very little point in trying to use these data for point scoring or as a basis for exhortation to improvement.

If firms, of whatever type, wherever they are, have both a competitive incentive and an economic rationale to use advanced technologies, they should and will usually do so; if not, they shouldn't and usually won't. Attempts by outsiders to incite or coerce firms to innovate before their circumstances are ripe will usually backfire.

7.1.1 FMI Research Survey 2007

The earliest survey referred to here was carried out in 2007 by the Construction Management Association of America (CMAA), in conjunction with FMI

110 BIM – the current state of play

Research, a management consultant specialising in the business of the construction industry.[1] The CMAA and FMI conduct an annual survey canvassing the views of construction clients on a variety of important issues of the day. In 2007 the focus of the survey was BIM and associated practices. Responses were obtained from about 200 of the largest public and private sector client organisations in the USA.

Considering how early it was carried out, the most surprising finding of this survey was that, in 2007, already 35 per cent of American clients had used BIM processes and technologies for a year or more. (By comparison, the 2010 NBS survey reported below found that fewer than about 10 per cent of UK client organisations were aware of, or using, a BIM approach.) The key features of these early US users were that they were predominantly private organisations, larger than average, national in geographic extent and that they carried out the larger projects. In short, they tended to be the more sophisticated clients who used more collaborative approaches and who placed greater reliance on formal information management methodologies. The most important finding about these organisations was that 74 per cent of them would be likely or very likely to recommend use of BIM systems. This message repeats throughout the surveys discussed here.

7.1.2 McGraw-Hill USA Surveys 2007–10

The Construction division of the US-based information services company McGraw-Hill (MHC) has carried out four important annual surveys of BIM adoption and usage in the US construction industry. Each of these surveys studied a different aspect of the BIM experience, as follows:

- 2007 Interoperability in the Construction Industry[2]
- 2008 Building Information Modelling (BIM):[3] Transforming Design and Construction to Achieve Greater Industry Productivity
- 2009 The Business Value of BIM:[4] Getting Building Information Modelling to the Bottom Line
- 2010 Green BIM:[5] How Building Information Modelling is Contributing to Green Design & Construction.

Amongst a broad number of other issues, in each of these surveys, BIM user firms across the industry were asked about their current and anticipated future

---

1 http://www.cmaanet.org/foundation-research-projects (retrieved 10 October 2010).
2 http://construction.ecnext.com/coms2/summary_0249-259123_ITM (retrieved 28 July 2010).
3 http://construction.ecnext.com/coms2/summary_0249-296182_ITM_analytics (retrieved 20 February 2010).
4 http://www.bim.construction.com/research/ (retrieved 4 October 2010).
5 http://construction.com/market_research/FreeReport/GreenBIM/ (retrieved 17 September 2010).

Table 7.1 McGraw-Hill reports: adoption intensity

Report Title / Year	McGraw-Hill Smartmarket Reports - Forecast and Actual Adoption Intensity								
	"Interoperability" / 2007					"BIM" / 2008		"Business Value" / 2009	
	2007 Estimate		2007 3 Year Forecast			2008 Actual	2009 F'Cast	2009 Actual	2011 F'Cast
Level of Adoption	2005	2006	2007	2008	2009	2008	2009	2009	2011
Architects < 16%	68	57	19	5	2	32	12	25	5
Architects > 16%, < 60%	18	20	29	34	39	25	34	38	28
Architects 60% +	14	23	52	61	61	43	54	37	67
Engineers < 16%	85	78	51	38	21	36	21	40	9
Engineers > 16%, < 60%	10	14	21	26	31	29	36	39	48
Engineers 60%+	6	8	27	36	48	35	43	21	43
Main Contractors < 16%	84	79	62	52	36	45	12	37	4
Main Contractors > 16%, <60%	10	12	18	22	26	32	50	42	53
Main Contractors 60% +	6	9	20	26	42	23	38	21	43
Owner / Clients < 16%	91	80	58	43	28	41	33	41	11
Owner / Clients > 16%, < 60%	6	16	27	38	48	18	21	41	47
Owner / Clients 60% +	3	4	15	19	24	41	46	18	42
Industry < 16%	82	73	47	32	15	38	18	34	6
Industry > 16%, < 60%	11	15	24	30	36	27	37	39	42
Industry 60% +	7	11	29	38	49	35	45	27	52
Note:	< 16% = Mainly Exploratory Activity. Low Actual BIM Usage.					> 16%, < 60% = Significant Commitment to BIM Usage			
	60% + = Largely, to Fully Committed to BIM								

Source: McGraw-Hill Smartmarket Reports: 2007, 2008, 2009.

levels of usage of BIM tools and techniques. The relevant survey questions for 2007–9 were similar and gave the results summarised in Table 7.1.

Note that these were the results for responding firms who were actually using BIM to some extent in the respective years. Thus, of all respondents, only 28 per cent of firms were actually using BIM in 2007. But this number had increased significantly – to 49 per cent – in 2009. (The 2010 report did not include comparable questions in this subject area.)

In both 2007 and 2008 users were asked to predict how intensively they expected to be using BIM by 2009. As the 'actual' usage level reported in the 2009 survey shows, forecasts for very low and very high levels of adoption intensity are generally not accurate. However, 2007 and 2008 forecasts for moderate to significant usage for 2009 were generally remarkably accurate. The surveys were not designed to extract strong data on these issues – as noted earlier, they were focused on the specific topics denoted by their titles. However, as the table shows, it would seem reasonable to observe that, at the significant commitment level of usage, firms are already using BIM fairly intensively, and are likely to continue to intensify their BIM usage in the immediate future – despite the dreadful economic state of the industry in recent years. So adoption levels are high and rising quite rapidly across all of the principal US industry user groups: architects, engineers, contractors and clients, and the intensity of usage is increasing significantly.

Amongst the four groupings, architects are both the most active and the most intensive users. Among BIM users, 37 per cent of architects were using BIM on 60 per cent or more of projects in 2009; 67 per cent expected to be doing so by 2011. The corresponding figures for structural engineers were 25 per cent and 50 per cent and for MEP engineers, 10 per cent and 40 per cent (2009 Report, p. 37). The reason for the differences between the design disciplines is generally thought to be because architects are already largely comfortable doing their own drafting on

CAD systems, so moving on to hands-on modelling is relatively straightforward. Structural engineers and, even more so, MEP engineers are less accustomed to doing their own drafting, so are less amenable to BIM modelling which, more so than CAD, requires the designer to create his or her own models.

A second reason why MEP engineers in particular are less advanced in their use of BIM systems is because the relevant tools are considered to be deficient in two broad respects. First, the available families of well-specified components doesn't adequately cover the scope of real-world mechanical and electrical systems, and those components that do exist lack attribute richness. Second, the process of importing and exporting components between modelling and analysis tools is cumbersome and inefficient, relying as it does on significant user intervention to determine the classes of objects and the particular information entities to be provided at each interchange between the different systems. Using this sort of fudged interoperability, it can be easier for designers to create their own, task-specific analysis models, rather than exchanging data between their models and project BIM models.

The range of tasks for which BIM is reported to have been used in these surveys is broad, but breaks down into three general areas:

- creating models – largely the process of building the 3D geometry (2008 Report, pp. 14–15);
- analysing models and model data – scheduling, structure, energy etc. (2008 Report, p. 37);
- viewing models – visualisation, clash checking, construction simulation etc.

The Green BIM 2010 report, as one would expect, elaborates on the current and future role played by BIM in the design of sustainable buildings and in their production – using carbon and energy efficient methods. A case study featuring Shanghai Tower, a super-tall building, highlights the way in which BIM models were used, to 'design the most efficient structural frame,' and in the search for existing high-quality products, so as to avoid the need to manufacture new customised products (2010 Report, p. 19).

A striking finding of all four surveys was the extent to which almost every aspect of the deployment of BIM becomes easier and more beneficial with experience. For example, both the ability to measure return on investment and actual return on investment are reported to improve rapidly as firms become increasingly expert in their implementation of the BIM approach and related techniques. More widely, amongst BIM users:

- 43% of experts see increased profits: versus 7% of beginners
- 77% of experts find reduced re-work: versus 23% of beginners
- 76% of experts produce better documentation: versus 26% of beginners
- 71% of experts find BIM helps win new work: versus 28% of beginners
- 61% of experts find BIM helps retain clients: versus 19% of beginners.

(2009 Report, p. 15)

A strikingly important remark was made by an architect on a second case study in the Green BIM report. He suggests that, contrary to received wisdom, BIM may reinforce or perhaps reinstate the central integrating role of the architect. Tom Chessum, principal in the firm of CO Architects, is quoted as saying:

> It puts us as architects in the position to guide (the team) by explaining to them what the overriding design goals and concepts were and to lead all that to fruition with their buy-in, as opposed to the old method of meeting the contractors after they ... have made all their own interpretation of our documents that weren't quite right, forcing us to have to defend the design.
> (2010 Report, p. 13).

7.1.3 Other surveys

McGraw-Hill carried out a second BIM survey in 2010, this time in Europe. The aim was to capture how firms in France, Germany and the UK were adopting BIM, and the benefits they were obtaining or hoped to derive from BIM implementation.[6] Broadly speaking the results were as might have been expected. In summary:

- US industry's adoption of BIM grew from 28% in 2007 to 49% in 2009: only 36% of the European market has adopted BIM
- Architects are the keenest adopters (47%), engineers (38%), contractors (24%)
- 45% of European users claim to be experts or advanced users
- 24% of European contractors are BIM users: versus 24% in the USA
- 34% of European users have more than 5 years' experience: versus 18% in the USA.

(BIM in Europe Report, p. 5)

Apart from the slight lag in implementation, there are no remarkable differences between the pattern of adoption in the USA and Europe. The 34 per cent of European users with greater experience than their American counterparts are probably those who have worked with products like Sonata/Reflex, and Graphisoft over the years. The UK experience of collaborative forms of project organisation will probably enable firms here to catch up quite rapidly in the move to a more generalised implementation of integrated forms of BIM-based project delivery.

A refreshingly frank, if slightly discordant note, is sounded in the case study of the University Campus Suffolk project in the BIM in Europe Report. Although the team did see some of the promised benefits of BIM, it was not a painless process. One can sense the effort involved, from the slightly forlorn comment of a senior architect on the job: 'We're not sure if the client will insist on it again, but it's good that we went through this process, because we expect

---

6 http://bim.construction.com/research/FreeReport/BIM_Europe/ (retrieved 28 October 2010).

114 BIM – the current state of play

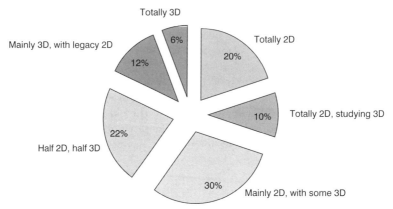

Figure 7.1 Distribution of 2D and 3D CAD usage, USA, 2010 (source: Robert Green)

there to be more clients requesting proper 3D in the future' (BIM in Europe Report, p. 43).

Figure 7.1 summarises a survey carried out and reported on in an article by Robert Green in *Cadalyst* magazine. As the experts stress, 3D CAD is not BIM. However, it would seem reasonable to suggest that the act of moving into 3D is a precursor to BIM. So, Green's figures suggest that about 40 per cent of the firms he canvassed have made serious progress in this direction by deploying 3D on at least half of their projects.

NBS, who produce the eponymous National Building Specification and other forms of information for the UK construction industry carried out a brief survey, also in 2010. The results were reported in a presentation by Dr Stephen Hamil, NBS's Head of BIM, and are available on the NBS website.[7]

Of about 400 responses, 40 per cent were architects, about 15 per cent each engineers and client personnel and about 2 per cent contractors. 43 per cent of this group professed themselves completely unaware of BIM, with only 13 per cent actively using BIM systems (compared – awkwardly – with MHC's 36 per cent European users at some identifiable level of usage). However, all of those aware of BIM expect its usage to grow significantly, reasonably quickly.

- 24 per cent of these expect their firms to use BIM for all or a majority of their projects in one year's time;
- 50 per cent of these expect their firms to use BIM for all or a majority of their projects in three years' time;
- 63 per cent of these expect their firms to use BIM for all or a majority of their projects in five years' time.

The benefits observed by NBS's BIM users were significant:

7   http://www.thenbs.com/bim/What_BIM_is_and_how_it_is_being_used.asp (retrieved 7 December 2010).

- Increased speed of delivery: 51% agreed
- Improved coordination of documentation: 81% agreed
- Increased efficiency of document retrieval: 84% agreed
- Improved visualisation: 85% agreed
- Increased profitability: 53% agreed
- Cost efficiencies: 61% agreed
- Changes to workflow and other practices: 88% agreed
- Adopted BIM successfully: 58% agreed
- Glad to have done it: 78% agreed.

Interestingly, where comparable issues were questioned, the NBS responses were slightly more positive than those reported in the McGraw-Hill surveys. In fact, given how early in the BIM adoption process the organisations in question are, these are strikingly positive results.

7.1.4 Conclusion

It almost goes without saying that the findings reported here should be read with caution. Individually, each of the surveys achieves its own objectives – captures some aspect or aspects of the BIM phenomenon – very well. To run them all together in an attempt to throw a more generalised light on the central subject is unfair to the surveys' authors, but with that caveat, it remains the case that the surveys covered here, individually but more convincingly, collectively, suggest quite strongly that:

- BIM is a real thing, being used with some degree of seriousness by upwards of a third of construction organisations in the USA and Europe.
- BIM usage is growing quite rapidly.
- Experienced users of BIM are deriving disproportionately more utility and benefit from its use than less experienced users.
- BIM increases efficiency and BIM improves profitability – the bottom line.

7.2 Case studies: introduction

Knowing that a significant proportion of organisations in the industry have started down the BIM path is encouraging. The benefits that user firms are enjoying seem to be broadly as the theory of BIM would suggest. In order to probe a little deeper, to capture a sense of how firms are actually using their BIM systems, a number of simple case studies were carried out. There was nothing scientific or even particularly systematic about the conduct of these studies. The aim was simply to flesh out slightly the survey statistics summarised in Section 7.1.

The five main vendors of parametric component-based AEC modelling systems: Autodesk, Bentley, Gehry Technologies, Graphisoft and Tekla, were each invited to nominate a firm or project which, in their view, was representative of current good practice in the use of BIM tools and techniques. The picture of each of the firms' use of BIM was based on a variety of sources, mainly discussions with named members of staff and with selected project team members from other firms as well as the firms' published project fact sheets.

Alan Baikie, Andrew Bellerby, Steve Jolley, Adam Matthews and Dennis Shelden all provided crucial help in selecting and recruiting candidate case study firms; for which many thanks.

7.3 Case study: Frank Gehry's architecture

If it had not been for BIM (not that it was called that, then), Frank Gehry would never have been able to create many of the remarkable buildings that characterise the past 20 or so years of his work. Gehry himself is indifferent to computers, but Jim Glymph, an early partner in his firm, developed a method of working which suits both Gehry's design style and also his philosophical approach to architecture. This was sketched out in an interview with John Tusa of the BBC in September 2005.[8]

Gehry acknowledges the influence of the modern movement on his work, but expresses discomfort with the stark coolness of conventional modernism; and he rejects the gratuitous ornamentation of post-modernist architecture. So he invented a whole new idiom. He started working on galleries and museums quite early in his career: relatively large buildings enclosing, usually, large public display areas; obvious candidates for the modernist treatment. A major concern of his from the beginning was to make these large spaces more approachable and human in scale and feel. He attempts to do this in two ways. First by breaking up the large areas into smaller, more intimate zones. And at the same time, while respecting and echoing quietly the elegance of the modern style, he tries to make it more attractive to human perception. He does this by careful distortion of the shape of the spaces.

In the Tusa interview, Gehry refers to some of the work of the painter Giorgio Morandi as being one source of inspiration. The painting to which this URL[9] leads is of a number of bottles and vase-shaped objects, clustered in a tight group. The explanation Gehry gives is that he tries to get a plan arrangement whose perimeter is conceptually similar to that which might be drawn if one took a pencil and scribed around the outside of the clustered group of shapes, then squash in the tops of the bottles slightly to generate some vertical interest. He stresses that his primary concern is to create interesting spaces inside his buildings; only when that is done does he get to work on the external treatment and the arrangement of the building in its physical location. It's not as clear cut as that however. There seems to be a continuous process of looping around in a simultaneous perception

---

8 http://www.bbc.co.uk/radio3/johntusainterview/pip/fmvd6/ (retrieved 10 November 2010).
9 http://upload.wikimedia.org/wikipedia/en/6/67/%27Natura_Morta%27%2C_oil_on_canvas_painting_by_Giorgio_Morandi%2C_1956%2C_private_collection.jpg (retrieved 14 November 2010).

BIM – the current state of play   117

Figure 7.2 Weisman Museum, Minneapolis (courtesy: Wikipedia)

of inside and out, space and surface, detail and overall, all of which has to be supported by his design tools.

The Weisman Museum in Minneapolis, Minnesota was Gehry's last major project to be documented primarily with drawings (Figure 7.2). Although, at first sight it looks similar to his more recent work: a jumbled-looking heap of intersecting shapes, the Weisman in fact comprises fairly regular, mostly conic forms which, though complicated in their arrangement, can actually be drawn using conventional plan, section and elevation views.

His later work, starting with the deceptively simple looking Fish sculpture at the Olympic Village in Barcelona, is all much more complex, involving extensive use of doubly curved surfaces (Figure 7.3). These vary between being extremely difficult and impossible to draw economically.

Glymph realised that to create buildings with such a degree of geometrical complexity he would have to make big changes to the way in which the practice supported Gehry's thought processes. In the first place, he saw that the only way to produce these surfaces would be to use the same sorts of systems as those used in car and aircraft manufacturing. So, in the early 1990s the firm started using CATIA, from Dassault Systèmes. This raised the problem of how to translate Gehry's innumerable sketches and physical models of all types into digital form. Initially this was done by physically measuring the models and transferring the relevant dimensions into the CATIA system; a tedious and error-prone process, similar to lofting in reverse. Other techniques included the use of coordinate

118 BIM – the current state of play

Figure 7.3 Gehry's Fish, Barcelona Olympic Village (courtesy: Wikipedia)

measurement machines (CMM), with three-axis articulating arms and sensitive probes, such as were used for early cranial scanning. More recently, the digitisation process has been carried out using laser scanning systems. Of course, Gehry himself still works in an utterly analogue fashion.

The second realisation that Glymph made was that having built the design in a CATIA model, he would also have to create the physical building using manufacturing techniques similar to those used to make the body and fuselage panels of cars and aircraft. He realised that conventional construction contracting arrangements would not work in this situation.

So, instead of working through a main contractor, he and Gehry sought out and made contact directly with manufacturers and fabricators who could take the numerical control (NC) data generated by CATIA and use it directly to control the machines used to make their products. They did this from the very beginning of Gehry's firm's becoming a digital practice. For example, the Barcelona Fish structure, which was part of the city's Olympic Village development project, was manufactured by the Italian cladding firm Permasteelisa, working directly with Gehry, rather than through the project's construction managers. The establishment of such direct relations with the component manufacturers on his projects is a fundamentally important aspect of Gehry's perception of his work, as an architect of the digital era, but fully in the *archi-tecton*, master-builder tradition.

As works of architecture, Frank Gehry's buildings are amongst the most interesting of the age. Structures like the Guggenheim Museum at Bilbao, (see Figures 7.4 and 7.5), are testimony to the fact that, as works of architecture, Frank Gehry's buildings are amongst the most interesting of the age. For the purposes of this book though, their most remarkable aspect is the way in which they are

BIM – the current state of play   119

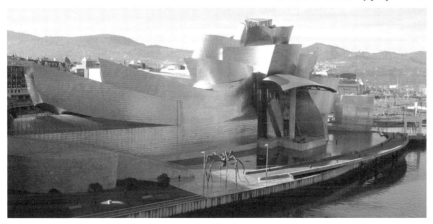

Figure 7.4 Guggenheim Museum, Bilbao (courtesy: Wikipedia)

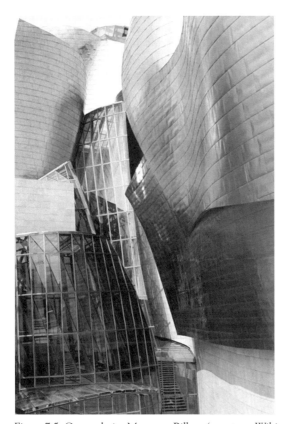

Figure 7.5 Guggenheim Museum, Bilbao (courtesy: Wikipedia)

designed and built. The key thing about these buildings is that they are constructed using effectively perfect information. Despite their extraordinary complexity, they are therefore completed on time, on budget and fully to their clients' requirements and expectations – surely the future of construction.

This case is based on the author's personal experience and on public domain information about Frank Gehry's practice and methodologies.

7.4 Case study: Ryder Architecture

Ryder Architecture was founded as Ryder and Yates in 1950, by Gordon Ryder and Peter Yates. Both men had worked previously with Berthold Lubetkin, a leading member of the Modern movement and both were heavily influenced by Modernism's key themes of design elegance and rooted social responsibility. The Modern approach can still be traced through the firm's work: rational, intelligent, functionally elegant buildings, which respect both their physical environments as well as the social and communal context in which they are situated.

The firm treats building technology seriously, both in terms of construction methods and of in-use building performance. This was reflected in its growing into a fully multi-disciplinary practice in the 1970s. As the engineering professions became increasingly specialised in subsequent years, the firm withdrew from engineering and scaled down its internal engineering team. Throughout however, Ryder has cultivated and refined the key skills required to build quick and flexible, but highly productive relationships with external design firms. So specialist groups in areas such as environmental and urban design are retained.

Although the founding generation is no longer involved in the practice, many of the current leaders of the firm have been with Ryder since the 1970s. So Ryder today displays clear continuity with the design ethos of the original Ryder and Yates. The firm today also maintains its founders' strong tradition of commercial acumen and operational competence in project delivery. The senior management of the firm take a direct and serious interest in the use of information technology, not only for the central design processes but also to help communicate ideas to clients and other project stakeholders, and to assist in technical coordination with other firms in their project teams.

The practice undertakes a wide variety of commercial and public-sector work, carried out under various forms of contract, including PFI, and two-stage and single-stage design and build, as well as more traditional forms. The firm currently comprises about 100 professional staff based in offices in Newcastle, Glasgow, Liverpool and London.

Like most practices of this type and size, Ryder are used to collaborating with a very wide variety of clients, design firms, contractors and product manufacturers. A recent survey of projects counted over a dozen active clients, ten contractors of various sizes, and 20 consulting firms with whom the firm is currently working actively, on live projects. Each of these collaborations involves the establishment of a complex combination of linkages between Ryder and the partner in question. As well as the organisational and social aspects of strategic and day-to-day, company-to-company relations, the linkages address the commercial relations between the two organisations: contracts, insurance and intellectual property issues and such like. The linkages also have to enable accurate, effective control of operational issues: internal and external communications, production scheduling, deliverables management, and so on. And, obviously, the connections between the partners must support the exchange of technical materials such as models, drawings, schedules and other types of documentation. The ability to set up and tear down these complex, multi-functional, project-specific connections quickly and cleanly is critically important. The interfacing functions must be thoroughly understood, and precisely specified – a remarkably sophisticated kind of organisational 'plug-n-play' in the business world.

The combination of ethos, attitudes and technical competence that motivates and underpins Ryder's working is illustrated clearly in the way in which they have taken to Building Information Modelling. Almost all of the key issues with BIM have surfaced on one or more of their projects: problems overcome; opportunities exploited.

Ryder's experience with BIM started in 2006, with two sixth-form centres for Redcar and Cleveland Council. The brief for the centres was to design facilities that did not feel institutional, encouraged participation and could support core skill and vocational route-ways for students. Development of a virtual building model supported engagement with end users and simplified the space planning process for the project team. A rendered animation in the form of a white model was also created to demonstrate spaces in a true context. This was updated with materials and textures based on briefing sessions with students.

The Redcar and Cleveland centres were followed in 2007 with an £18.5m student accommodation project that was running out of time (Figure 7.6). The key bottleneck was seen as being geometrical coordination amongst the five-firm design team. The solution adopted was to accelerate the work using Autodesk Revit as a design coordination tool. That worked. And many other benefits quickly emerged. For example, although it was late in the day, by integrating the architectural model with a topographical model that had been created some time earlier, it was possible to carry out a number of environmental optimisation exercises, including reviews of the orientation of the buildings on the site. The model was linked to NBS specification software, which helped ensure compliance with building regulations, and it was also used to generate output data files which were passed directly to the manufacturers of the 450 bathroom pods used in the buildings.

A £42m fire station on a tight site in south London, posed a different set of challenges. The client set high environmental standards, requiring that the building – a fire station, remember – should achieve BREEAM Excellent as a minimum. There were also severe problems of overshadowing and space constriction with adjacent structures. The building was modelled in detail by each of the three main design partners, and architecture, structure and MEP models were coordinated through regular FTP file exchanges. The construction

Figure 7.6 Victoria Hall student accommodation – 3D BIM model (courtesy: Ryder Architecture)

programme was simulated in the model, by integrating data from the project planning system. This was used to test and prove that, given the tight budget on which the project was based, a prefabricated panelised brickwork system would be preferable to *in situ* brickwork.

The renovation of Manchester Central Library, a Grade II* listed building, was carried out as part of the wider refurbishment of Manchester's Town Hall complex (Figure 7.7). All of the consultants who made up the core project team were experienced in the use of BIM techniques. The team decided early in the project to commission a Revit survey model using high definition laser scanning technology. This enabled the complex geometry of the building to be captured very accurately, and saved significant effort in the production of detailed design information in Stage D.

Early in their use of BIM, Ryder found that the actual modelling tool was only part of the solution – BIM is actually a whole approach to the technical aspects of managing and developing the design of a building. As noted above, the firm has long experience of working with others on its projects, so skills in basic CAD data exchange are well developed. To a great extent, these are the same as the skills required for BIM, so only limited adaptation is required in the creation and management of the data exchange files. However, to obtain best advantage from BIM interoperability requires a more structured approach, in the sense that the firms need to coordinate their information interchange plans from the outset. Ryder do this through the use of a BIM execution plan for each project. This is developed collaboratively with the other designers and

BIM – the current state of play   123

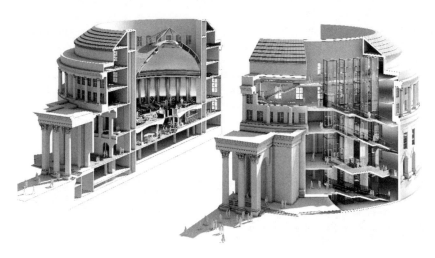

Figure 7.7 Manchester Central Library – visualisation derived from BIM model (courtesy: Ryder Architecture)

documents the responsibilities, technologies, naming and file format standards to be used. The execution plan must obviously be agreed at the very beginning of the project.

Ryder deploys internet-based information exchange technology, as well as their proven FTP file transfer facilities. On a recent £26m schools project, the firm collaborated with an outsourcing partner, Eigen, based in India, who developed some production packages for the Revit model, under Ryder's supervision.

So, in one way or another, Ryder Architecture has explored a very wide range of the potential areas of benefit that might result from the use of the BIM approach on their projects. It is particularly interesting to see how the firm has folded BIM usage, pretty well seamlessly and supportively, into the traditions of the practice.

A final, broader point, this time regarding a more recent building – the £25m Grimsby University Centre (Figure 7.8) – is worth noting briefly. Ryder are acting as architect and design team leader on this project, working with AECOM, who are providing full civil, structural and MEP design services. Both firms are working in BIM on the project and both are well experienced in its use, with the result that the tender documentation is being generated from a fully coordinated, composite model. This means that the tender documents are being prepared on the basis of more or less perfect design information. The bidders' responses are proving to be accurate in their interpretation of the scope of work, and are closely competitive, as this book's theory suggests they should be.

Thanks to Nahim Iqbal and Richard Wise – Ryder Architecture.

Figure 7.8 Grimsby University Centre – multi-disciplinary model (courtesy: Ryder Architecture)

7.5 Case study: Ramboll

Ramboll's UK division started life as Whitby and Bird, founded by Mark Whitby and Bryn Bird in 1983, and joined by Mike Crane in 1985. The firm acquired a reputation for their thoughtful approach to the creation of buildings and other structures; a sort of engineered architecture, fusing architecture and structure with the environmental design of the building. Rather like Ryder Architecture, they combined a clear design ethos with business competence and a strong commitment to systematic and effective project delivery methods.

Whitbybird, as it was then known – about 650 people strong – merged with the global firm Ramboll in 2007 to form Ramboll Whitbybird – simplified to just Ramboll in 2009. Ramboll, which was originally founded in Denmark in 1945, has been involved in some of the largest projects in Europe, including both the Great Belt Fixed Link and the Øresund Bridge which together provide a continuous road link between mainland Europe and Scandinavia, via Denmark. Ramboll in the UK retains characteristics of both its component parts: solid engineering informed by flair and innovation.

The recently completed Norwich Open Academy was a £20m lump sum design and build project for Norfolk County Council (Figure 7.9). The main contractor, working under an Academies framework contract, was Kier Eastern. Architects were Sheppard Robson, Ramboll was the structural engineer, and WSP provided geotechnical, M&E and fire engineering design.

The design strategy for the building reflects strongly the school's specialism in environment and engineering. The main building comprises a three-storey Open

Figure 7.9 Norwich Open Academy (courtesy: Kier)

Forum amphitheatre, around which circulation and classrooms for 950 students are arranged. The structure of the building comprises a honeycomb of five-ply, cross-laminated timber wall and floor panels. A total of 3,500 m^3 of these panels were fabricated by the Austrian company, KLH Massivholz, shipped to site on large trucks, on a just-in-time basis, and lifted directly into position from the beds of the trucks, using mobile cranes. Mobile access platforms and deck edge protection barriers were used to speed up the work and to avoid the need for scaffolding.

The advantages of this form of construction are numerous and significant. First, because the structure is so much lighter than a conventional steel or concrete equivalent would have been, the requirement for foundations and associated groundworks were greatly reduced; to the extent, for example, that all of the spoil generated in excavations could be retained on site. Second, and again by comparison with steel or concrete equivalents, the carbon footprint of the building was substantially less than it would otherwise have been. The embodied carbon in the timber frame alone was 60 per cent less than would be possible with conventional materials. The timber panel manufacturing process is zero waste; on-site waste was almost as low. Erection of the structure was very rapid, and follow-on trades had early, clean access to their workfaces. (Almost no wet trades were required internally.) Hangers and other surface fixings were easily attached to the timber structure – saving substantial amounts of effort and time.[10] The site incurred no lost time incidents and the total number of structure-related RFIs was ten, both achievements being largely attributable to the form of construction

---

10 http://www.kier.co.uk/strategic_alliances/projects_details.asp?p=593&x=&co=25 (retrieved 28 Jan 2011).

used. John Claydon, Kier's contract manager, stated that '... taking everything into account, this approach, at least for buildings of this type and scale, about breaks even with conventional methods'.[11]

A crucial feature of this form of construction is that, although as noted, it is easy to fix to the surface of the timber panels, it is very undesirable to cut holes in them on site. This means that the location and sizes of all services routes and of all windows and doors and other applicable architectural features must be fully determined before fabrication commences, so that the necessary penetrations and cut-outs can be built into the relevant panels during the process of manufacture. In order to do this Ramboll set up a workshop-based, inter-disciplinary coordination process, using their Bentley 3D model as the primary reference model. Drawings of the other disciplines were imported into the Ramboll model and checked for coordination. Clashes and other problems were identified and corrective actions agreed. At a sequence of these workshops, over a relatively short period, the team worked its way, floor by floor, through the building until all areas had been approved and signed off. The complete, coordinated model and associated drawing files were then passed over to KLH who used them as the basis of their own production model, to generate the data streams required to drive their CNC machinery.

A second example of the Ramboll approach to integrated engineering design is provided by the firm's work on the new Hepworth Gallery, part of a redevelopment of the waterfront area of the city of Wakefield, where Barbara Hepworth, the sculptor, was born (Figures 7.10 to 7.12). The project budget was about £22m. The architect was David Chipperfield Associates (DCA), the main contractor was Laing O'Rourke Northern; Ramboll provided all engineering design services.

The design objectives included the creation of interesting, large spaces within which to display a variety of works of art, including examples of Hepworth's sculptures. While the architecture was intended to have a substantial presence of its own, it was required not to compete in the minds of viewers with the works on display, but rather to support and illuminate them. The solution is a flowing succession of elegant spaces, receptive surfaces and precise, engineered lines, with delicate and complete control of lighting and other aspects of the internal environment. It almost goes without saying that the building was also required to be highly energy efficient, with a minimal carbon footprint.

The overall external assembly of shapes and spaces is intended to reflect the jumble of small industrial buildings that originally occupied the site. The arrangement of the building blocks, the colour of the concrete walls and their 'as struck' finish also echo some of Hepworth's sculptures. The building's location at the end of a peninsula, and effectively in the river, added somewhat to the design and construction challenge.

The technical systems issues involved in the design of such a complex and finely detailed building as this are formidable. The different Ramboll disciplines involved all use their own specialist design software packages; some of these are Bentley based, others, Autodesk. And, of course, the Ramboll team must exchange large

---

11 Private communication, 17 February 2011.

BIM – the current state of play   127

Figure 7.10 Wakefield Waterfront – before (courtesy: Wakefield Council)

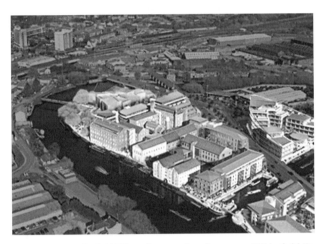

Figure 7.11 Wakefield Waterfront – after (courtesy: Wakefield Council)

Figure 7.12 The Hepworth Gallery, Wakefield (courtesy: Ramboll, Jonty Wilde)

volumes of design information with the other members of the larger project team, in an efficient and controlled manner.

A major part of the firm's response to this challenge was to create a complete, multi-disciplinary 3D model of the building. This model incorporated all relevant aspects of the DCA design as well as the work of all of the Ramboll disciplines working on the project. The procedures, CAD standards, and other aspects of protocol required to support this sort of communciations and information interchange – all based on Ramboll's established procedures – were agreed in a series of technical workshops at the beginning of the project. The result in this case was a well-coordinated design, accurate construction documentation, and relatively few field clashes.

Other, internal benefits that Ramboll gained from their BIM-type approach on the project included early support for value engineering exercises, easy generation of visualisation materials and a significant time saving over conventional CAD drafting methods. Ramboll have demonstrated that this integrated 2D and 3D, BIM-based approach offers an average saving of 20 per cent of CAD technicians' time per project.

Thanks to Steve Wright – Ramboll, John Claydon – Kier, and Terry Hughes – Laing O'Rourke.

7.6 Case study: Team Homes Limited, Parmiter Street development

Team Homes is a contractor/developer whose mission is to design and build high-quality, affordable housing using modern methods of building design and construction. The company's £25m Parmiter Street development for Family Mosaic Housing Association in the East End of London (Figure 7.13), contains 105 three-, four- and five-bedroom family homes on a tight inner-city site. The scheme provides a mix of shared ownership and rental units.

Working with John Robertson Architects (JRA), Hemsley Orrell Partnership structural engineers and Max Fordham mechanical and electrical engineers, Team designed the development to meet Code 4 of the Code for Sustainable Homes. A CHP system, photovoltaic solar panels and high levels of insulation contribute to achieving the standard. The scheme also attained Lifetime Homes and Secured by Design standards.

The Parmiter Street scheme was designed to ensure that 65 per cent of the site area is given over to amenity space. The homes feature front gardens, balconies, raised private terraces and private roof gardens. There are communal gardens, children's play areas and public spaces. And on top of that (literally) there are rooftop allotments where tenants can grow their own fruit and vegetables. The scheme consists of medium-rise blocks formed around two landscaped courtyards at first floor level (residents car parking being provided below). There are three six-storey blocks of apartments running north–south linked by four-storey maisonettes on the north side and three-storey town houses on the south.

BIM – the current state of play   129

Figure 7.13 Team Homes, Parmiter Street development (courtesy: John Robertson Associates)

Team adopted a firm, hands-on approach in the management of the design phase, with the design team members being co-located at Team's offices. Team also persuaded the consultants to collaborate in the development and deployment of a Graphisoft, ArchiCad BIM model. JRA created and managed the master reference model. They also initiated and maintained the BIM protocol, but of course the other members of the team were involved in regular reviews and updates of this document as the project developed.

Team were particularly keen to use Graphisoft's virtual construction (Vico) software, for quantity take off and construction simulation. In order to do this every object and element in the model had to be coded to carry the necessary cost and programme information. For the Parmiter Street project a single Graphisoft user was designated model manager, with authority to create new components; anyone who needed additional components was required to have them developed by the model manager.

The scheme was modelled to RIBA Stage D by the architects alone. Among other applications at this stage, the model was used to evaluate between the proposed heavyweight construction system and a more conventional *in-situ* frame with lightweight cladding. An important part of this exercise was the comparison of the energy performance of the competing proposals. This was performed by exporting files in gbXML format from the ArchiCad model to an analysis package called DesignBuilder, which incorporates a high-speed simulation processor called DBSim. The early model was also used for visualisations which were prepared to support the planning application (Figure 7.14).

130 BIM – the current state of play

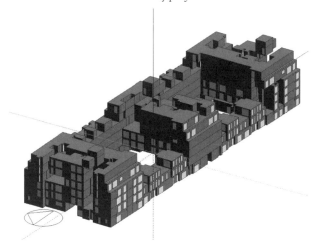

Figure 7.14 Parmiter Street – energy performance model (courtesy: John Robertson Associates)

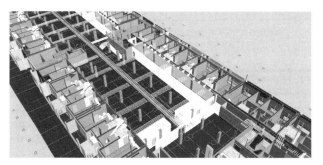

Figure 7.15 Parmiter Street – coordination model (courtesy: John Robertson Associates)

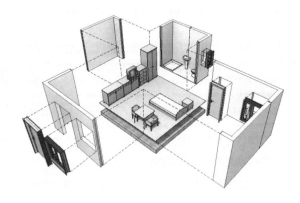

Figure 7.16 Parmiter Street – precast wall panels (courtesy: John Robertson Associates)

Production information (RIBA work stages E–G) was a full team effort. Generally the level of detail modelled for work stages E–G was that which could be seen when printed out at 1:50. However, in some cases this was exceeded when a particular detail warranted a greater level of attention. During this phase the model was used intensively in clash detection and design coordination workshops (Figure 7.15). Sub-contractor procurement packages included quantities and schedules taken from the model.

The scheme is constructed using an 'L'-shaped precast structural concrete panel system, with precast beam floor slabs (Figure 7.16). Parametric 3D library parts were developed for these key repeating elements to facilitate the design and speed up the programme. The library part for the 'L' panels allowed for the addition of window, door and services openings in the panel and whilst each leg of the panel could be up to 6 m long, the part would not allow the user to extend the legs so that the crane weight on site would be exceeded (calculated by the volume of concrete of each panel). Neither could a user make a leg less than the 600 mm minimum norm. Each library part when placed generated its own unique ID number.

The structural engineer produced schedules and a fabrication sheet for each panel generated from the BIM. These details were passed to Team's fabrication yard, where the panels were cast. Using this system allowed the building to become watertight at an early date and took the rain screen cladding off the construction critical path. It also ensured that openings in the structural panels were located exactly correctly, so that doors, windows, block-outs and services runs all fitted on site, precisely as designed in the model (Figure 7.17).

Thanks to Ben Wallbank.

Figure 7.17 Parmiter Street – wall panel installation (courtesy: Team Homes, John Robertson Associates)

132 BIM – the current state of play

7.7 Case study: Llanelli Scarlets Rugby Stadium – Parc y Scarlets

In late 2005, Costain were engaged in an early contractor involvement (ECI) role by Carmarthenshire County Council, to carry out remediation to some old mine workings at Pemberton, outside Llanelli (Figure 7.18). Costain's lead consultant was URS. Amongst other elements, the project included construction of a new 15,000-seat stadium for Scarlets RFC, formerly Llanelli RFC. Costain awarded a design and build contract for the new stadium project to Andrew Scott Ltd, part of the Rowecord Group, in September 2007. The contract value was £20.3m, with an agreed duration of 66 weeks. The design was led by Miller Partnership, a Glasgow-based architectural firm, with particular expertise in stadium design. Structural engineering was managed and coordinated by Rowecord; URS provided MEP and other design services. Steelwork fabrication and erection was by Rowecord, for whom Edge Structures provided design services and 3DS Limited provided drawing office services.

The short project schedule required a very aggressive design programme, particularly for the structural elements of the stadium. The design team agreed that this could best be achieved by sharing their information using a 3D structural BIM reference model. As Rowecord and 3DS both use Tekla Structures software, it was agreed that Edge would use Tekla and that the reference model would be developed in Tekla. A simple strategy for developing the BIM model was agreed during the early design team meetings. This included agreement on file formats, naming conventions and a basic workflow, whereby individual sub-models could be exchanged fairly easily and confidently amongst the team members.

The initial structural model was developed by the Rowecord and Miller teams working closely together to develop the overall stadium geometry, sight lines, grid systems, circulation, levels and suchlike. Miller work with Bentley software, so they created their architectural model in MicroStation and exported views of this to Rowecord in dgn format. The exports were worked up into the Tekla model by the Rowecord team, and the results were fed back to Miller, who used a Tekla web viewer to review and comment on the evolving structure. This intense,

Figure 7.18 Pemberton remediation site (courtesy: ©URS Corp/David Lawrence)

BIM – the current state of play   133

continuous, looping file exchange process moved from section to section of the structure and continued until all four sides and the corner sections of the stadium were complete.

This approach enabled Edge to carry out the analysis, member sizing and connection detailing section by section, freezing off each section progressively. The frozen material was issued as drawings to the other members of the design team and, together with the relevant member schedules, to Rowecord for fabrication planning. Edge's Tekla design model files were issued to 3DS for fabrication detailing. All fabrication drawings were generated directly from the Tekla model which was also used to produce the numerical control data files needed to drive the automated steel handling, cutting and drilling machines in the Rowecord works.

3DS took the design model and added the necessary fabrication-level connection detailing as well as the details of cladding rails and purlins. The terrace seating structure was built of *precast* concrete sections from Bison Manufacturing. The design of this structure was developed in precise detail, collaboratively between Bison, Edge Structures and 3DS. In order to provide good sight lines from all seating positions, the side terraces are slightly curved, as Figure 7.19 shows. Although the plank design was rationalised as far as possible, the curved layout necessitated the creation of over 400 different plank shapes. The 3DS team included the details of the entire precast design in the Tekla model to ensure that

Figure 7.19 Precast concrete terraces (courtesy: Rowecord Engineering Ltd, PJA Video)

134   BIM – the current state of play

Figure 7.20 The new Parc y Scarlets (courtesy: Rowecord Engineering Ltd, PJA Video)

each of the terrace planks sat correctly on its location point and that all other precast elements – slabs, walls and entrance/exit ways – were correctly situated on the supporting steel structure.

Throughout the detailed design, manufacture and construction phases of the project, the Edge Structures model was used as the reference model. All queries and other issues were evaluated against this model and necessary amendments were circulated to the other partners by means of Tekla files, web viewer files and other formats generated from the model. The structure took 20 weeks to erect on site. There were, literally, no field RFIs. The stadium went up 'like a giant Meccano kit', as Scott's site manager described it (Figure 7.20).

Thanks to Paul Benwell – Rowecord Engineering Ltd, and Jason Gething – Edge Structures.

7.8 Conclusions

These case studies, brief though they are, capture most of the key issues concerning the deployment of BIM systems and BIM methods in today's industry. First, it must be said that none of the cases represents an implementation of the true, ideal BIM approach. Recall from Chapter 6 the definition of BIM as comprising one or more parametric component-based modelling systems sharing information using neutral data exchange standards and working to agreed, project-wide protocols. The models used in these cases are typically hybrids of BIM and conventional

CAD; the data exchanges typically use proprietary file formats; and the exchanges are generally undertaken in a fairly *ad hoc* manner, rather than according to agreed protocols. So these might be called early BIM projects; not quite conforming to the purist's standards perhaps, but they display the main characteristics of true BIM operations, and they generate many of the benefits that one would expect of 'full' BIM.

Things are not uniformly positive in the process of diffusion of the BIM approach, however. Although this is not generally the case with the projects described here, it appears from anecdote and discussion in the industry that a number of influential design firms are expressing reluctance to share their models, particularly with contractors. The reasons given are usually to do with issues of professional liability and intellectual property and sometimes fee reimbursement. It also seems that some specialists, notably steel fabricators, have started to create their own fabrication models, from consultants' 2D drawings, rather than use the designers' models.

Both of these types of resistance are understandable, given the current, early state of development of the BIM approach. However, the reasons why they arise need to be explored and properly understood. And systems vendors and standards bodies must take them into account more explicitly than currently seems to be the case.

The overall message that one can take from this chapter is that the BIM approach is gradually becoming part of the mainstream of construction industry operations. Firms in the industry are definitely becoming more comfortable and more expert in their deployment of BIM systems. However, as the next chapter makes clear, the interesting strategic question is not really so much what the industry can do with BIM, as what BIM will do to the industry.

8 IT usage in construction and other industries

8.0 Introduction

The central argument of this book is that the two main problems of the construction industry are its persistently low profitability and its chronic failure to deliver projects predictably. All of the other problems of the industry – its low levels of investment and R&D, its poor safety record, its careless environmental impact, its apparent disregard for human capital, its poor record of customer satisfaction, its poor social status and low self-esteem – can, in one way or another, be traced back to these two. If these two fundamental problems could be solved, many of the industry's other problems would fall away.

The argument goes further, to suggest that the main underlying cause of both of the two key problems is the industry's persistence in trying to organise hugely complicated projects in the production of some of the most complex objects devised by the human mind, using primitively low-quality information, exchanged using correspondingly cumbersome and unreliable communications techniques and technologies. The main purpose of this book is to show how the new(ish) technical approach called Building Information Modelling might change the industry fundamentally for the better by dramatically improving the quality of information and the methods of communication used in construction.

In the book so far, the discussion of these issues has been presented in the isolated context of the structures and modes of operation of the construction industry. But construction is not alone in this. Almost all other sectors of the modern economy have already been transformed as a result of improved information management systems and techniques.

The purpose of this chapter is to place construction in that wider context. The aim is not to force the industry into some generic analytical mould; it's not to suggest that whatever is good for manufacturing or banking or retail, should also be good for construction. But by looking at the particular way in which game-changing IT was deployed in each of these sectors separately, we can perhaps draw out some of the fundamental issues involved in IT-driven transformation, and subsequently consider how those issues might play out in the construction industry.

This chapter opens with a review of how other industries have incorporated computers and digital communications into their operations. In his major work *The Digital Hand*, James Cortada, a leading historian of modern technology, describes how the work of 80 per cent of the US economy has been radically transformed through the implementation of information technologies. *The Digital Hand* takes the form of three large volumes, totalling over 2,000 pages of close observation and informed discussion. It covers almost every significant sector of the economy. But the construction industry, at 8–10 per cent, arguably the largest discrete sector of all modern economies, is not once mentioned. The second part of this chapter discusses why this might be.

8.1 The digital revolution – changing the nature of work

James Cortada, in his trilogy of books, *The Digital Hand*, suggests that economic progress in the West has evolved through three distinct, revolutionary phases:

- First there was the industrial revolution, involving primarily the concentration of manufacturing production in factories, and a reliance on Adam Smith's Invisible Hand of the market and enlightened self-interest to guide the operation of the economy.
- Next came the managerial revolution, driven primarily by the economies of scale achievable by very large, complex, integrated manufacturing organisations. This form of economic organisation was guided by the power and capability of professional managers – A.D. Chandler's Visible Hand – exercising explicit control over the flows of goods and services.
- Finally, and still playing out, he identifies the digital revolution, in which any process that can be programmed will be programmed, if doing so increases productivity or reduces costs in some other way. In the digital revolution, control over economic production is achieved using computers and other forms of digital systems – the Digital Hand.[1]

Cortada's *Digital Hand* trilogy examines the experience of 19 major industries, as well as government and other public bodies in the United States. In total, these account for 80 per cent of US GDP. In every case he identifies at least one information technology application that he describes as having transformed the area of activity in question. Every one of these has undergone IT-driven change that has been described by economic and social commentators as 'transformational', 'revolutionary' or 'explosive'.[2]

---

1 Cortada, J.W., *The Digital Hand, Volume 1: How Computers Changed the Work of American Manufacturing, Transportation, and Retail Industries*. Oxford: Oxford University Press, 2004.
2 Cortada, J.W., *The Digital Hand, Volume 2: How Computers Changed the Work of American Financial, Telecommunications, Media, and Entertainment Industries*. Oxford: Oxford University Press, 2006, p. 151.

Cortada leaves the reader in no doubt that digital technologies are transforming industries and entire economies so profoundly that the process can justifiably be called revolutionary. However, he is careful repeatedly to stress that this revolution is not happening in a social, historical or organisational vacuum.[3] The Invisible Hand of market forces, the Visible Hand of rational economic management and more recently the Digital Hand, are key stages in the continuous evolution of economic production methods. The process is driven by innovation in a competitive environment. It is as natural as Darwinian selection in the biological world, and no more stoppable; the survival of the economically fittest under price competition in free markets.

The crucial requirement for innovation to take place is the presence of effective competitive pressure. A central argument of this book is that construction has managed to avoid the pressure of competition and has thereby avoided the need to innovate.

8.2 The diffusion of innovations

A large proportion of the US economy has undergone substantial technology-driven change in its mode of operation and structure over the past 50 or so years. Sections 8.3 and 8.4 will review this experience in an attempt to illuminate the combination of opportunity and challenge that BIM represents for construction today. The main questions to be asked in regard to the adoption of any innovation are:

- What is the real nature of the innovation at hand?
- How rapid is the adoption process?
- How are the promised benefits to be measured at the outset?
- Who benefits, who loses?
- Who supports, who obstructs implementation?
- How can the level of adoption be assessed at any point in time?
- How can actual benefits measured after the fact?

E.M. Rogers has provided an analytical approach that is both straightforward to work with and useful in exploring most of these issues. Figure 8.1 illustrates some of the key features of Rogers' analysis: the 'S' shaped adoption curve; the bell-shaped distribution of adopter attitudes, and the point of critical mass in the adoption process, beyond which the diffusion of the innovation will be self-sustaining.[4]

Rogers[5] defines innovation diffusion as being '… the process by which an innovation is communicated through certain channels, over time, amongst the members of a social system'. Each of these four main elements breaks down into sub-issues, as follows.

---

3 Cortada (2006), p. 478.
4 Rogers, E.M., *Diffusion of Innovations*. New York: The Free Press, 2003, pp. 11, 281.
5 Ibid., pp. 12–38.

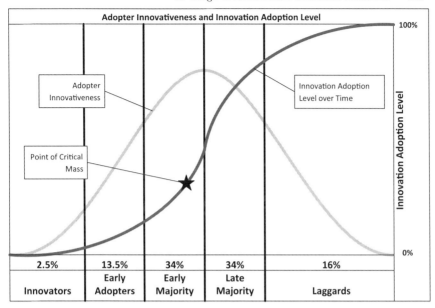

Figure 8.1 Rogers' innovation diffusion model (source: Rogers, *Diffusion of Innovations*)

8.2.1 The innovation itself

This is usually a combination of a hardware tool and the software, or knowledge, required to use it. The relevant attributes of the innovation are:

- relative advantage: the degree to which the innovation improves on the idea it supersedes;
- compatibility with existing values, technologies and needs;
- complexity: its ease of comprehension and implementation;
- trialability: the extent to which it can be experimented with, before implementation;
- observability: the ease with which the results of an innovation can be observed by others.

8.2.2 Communications channels

This refers to the means by which information about the innovation is passed between individuals. They include mass and specialist media, interpersonal communications and technical channels like the internet.

8.2.3 Time

The time required to complete the decision-making process. The individual needs time in which to make his or her decision: time required to learn about the innovation; time in which to form an opinion about it; time to actually make the decision and then time to implement and confirm the decision to adopt the innovation. Two additional time-related issues are the readiness – earliness or lateness – of the individual to adopt the innovation, as illustrated by the adopter categories in the distribution diagram in Figure 8.1: innovators, early adopters, early majority, late majority, and laggards. The final, time-related consideration is the actual rate at which the innovation is adopted by the individual's community or system, illustrated by the S curve in Figure 8.1.

8.2.4 Social system

The social system is the context of the interconnected units involved in the innovation decision-making process: individual members of a community or colleagues in an organisation, for example. Considerations here include such issues as the social and communications structures within the adoption community: system norms; the presence of opinion leaders, communications networks and change agents. It also includes consideration of the types of innovation decision being made, whether the innovation is or might be considered optional, collective, or imposed by an authority of some sort, or by a few powerful members of the system.

Rogers' four-element framework is used here as a useful reference checklist, a background for the discussion of the innovations addressed in the following sections.

8.3 General patterns of IT adoption in industry

This book is concerned with the impact of information technologies on business organisations, their employees, and their operational processes. So scientific and military systems are outside its scope. So largely, are administrative systems, defined for convenience as including accounting, billing, payments, and general office automation systems. These are all useful and important systems, obviously, but their impact has, by and large, been relatively neutral. The main focus here is on the primary, value-creating, operational activities of companies and the impacts of IT in these areas, which is where true transformation has occurred. Under the present analysis, technological innovation proceeds in four phases, as follows.

8.3.1 Phase 1: Early years – the search for productivity

As Cortada points out in *The Digital Hand* and more extensively in *Before the Computer*,[6] automatic, generally mechanical, methods of managing information have been in use in modern economies for well over a century. These include typewriters, calculating machines, punch card machines, cash registers and the

---

6 Cortada, J.W., *Before the Computer: IBM, NCR, Burroughs and Remington Rand and the Industry They Created 1865–1956*. Princeton, NJ: Princeton University Press, 2000.

telegraph. The application of these devices intensified significantly as growth, in the American economy particularly, accelerated during the inter-war period, and volumes of business information mushroomed.

In parallel with these developments in information management, a second area where increasing mechanisation occurred was with the introduction of increasingly sophisticated factory machines and systems of machines in manufacturing industry. The immediate aim of automation in this case was to augment or substitute for human labour and dexterity with machines. Increased productivity, higher quality and lower cost were the ultimate goals.

So innovative deployment of automation and mechanisation should be seen as part of a long historical process. Developments of the past 50 to 60 years, which from today's vantage point may seem dramatic and rapid, should more realistically be seen as part of this slower, longer term picture that in key respects is still unfolding.

A second point to note about technological innovations prior to about the 1970s (it varied from industry to industry) was that they tended to address single-point applications. Thus manufacturers invested in individual new machines, or relatively simple systems of machines. And firms tended to purchase stand-alone software applications to support individual business processes, like procurement, materials management, billing and so on.

The important feature of these initiatives was that they could generally be taken at a relatively local level by managers who were directly responsible for the processes in question. Being experts in the processes, they could evaluate the attributes of the innovations, as suggested by Rogers – relative advantage, compatibility, complexity, trialability, and observability – with confidence. So the investment was both non-strategic and relatively low risk.

Managers working at the operational level in their companies were also more likely than senior executives to keep abreast of technical developments in their specialist operational fields. They did this either through professional publications, the relevant trade press, trade shows and conferences, and simply through contact with their peers in the industry, all again as Rogers suggests they should.

These considerations were hugely important. In all of the industries that he reviews, Cortada describes the large majority of American managers as being fundamentally timid, conservative, and resistant to change. Far from being the men who tamed the West, the people Cortada describes had more or less to be dragged into the new technical era and new ways of doing business: 'They moved to embrace computing ... only when the case for them proved compelling or circumstances forced reluctant management teams to sharpen their productivity.'[7]

There is no doubt that the patterns of innovation diffusion that Rogers identifies apply particularly to this relatively early period in the digital era in the USA. In a social system whose norms include the most intense levels of competition, the influence of relatively few innovators and early adopters is strong. And, given the availability of very effective communications channels,

---

7   Cortada (2004), p. 158.

it is hardly surprising that these early sorts of single-point innovations were adopted very rapidly.

8.3.2 Phase 2: Middle period – local optimisation

One of the problems with the stand-alone solutions of this early period was that, although they might improve product quality and increase the productivity or efficiency of the particular process to which they were applied, it was often not clear that they reduced the overall cost of the firm's operations, or led to significant overall productivity gains. The new process or new machine might make an individual function more efficient, but this might have knock-on effects elsewhere in the organisation that offset the local improvement. In this situation the firm's overall productivity – output per man-hour – does not rise because, for example, as more and more information is generated by these new machines and systems, the tendency is to add more and more people to handle all the new information.

This is a dangerous situation. Innovation decisions are effectively made by functional managers who are unable to see the big picture. They are signed off by higher-level executives who are unable to understand the detail of areas outside their career zones of expertise. And they are supported by technology vendors and their attendant 'implementation' consultants who don't know the business well enough. This pattern of behaviour contributed significantly to what was called the 'productivity paradox' in the US economy. As Brynjolfsson described it in 1993: '… delivered computing power in the US economy has increased by more than two orders of magnitude since 1970, yet productivity, especially in the service sector seems to have stagnated'.[8]

This was an area of quite a heated debate during the early 1990s. The search for answers triggered the development of a number of management 'theories', the hottest of which was probably business process re-engineering (BPR). Thomas Davenport, one of the gurus of BPR, was amongst a number of writers who professed deep scepticism about the economic benefits being generated by companies' investments in IT. His argument was that deployment of innovative IT had to be done hand-in-hand with a fundamental re-design of the related business processes: 'Managers seeking returns on IT investments must strive to ensure that process changes are realised. If nothing changes about the way work is done and the role of IT is simply to automate an existing process, economic benefits are likely to be minimal.'[9]

This approach had two effects. First, re-engineering was used by firms as a cover for savage culling of the ranks of middle management. Given that middle management is where most of the useful corporate knowledge resides in most companies, this amounted to a sort of crude corporate lobotomy. Secondly, it

---

8 Brynjolfsson, E., 'The Productivity Paradox of Information Technology', *Communications of the ACM.*, 1993, 36(12): 66–77. Also, Cortada, 2004, pp.35–40. More pessimistically: Roach, S.S., 'Economic Perspectives', *Morgan Stanley*, January 1991, pp. 6–19.
9 Davenport, T.H., *Process Innovation: Re-engineering Work Through Information Technology.* Cambridge, MA: Harvard Business School Press, 1993, p. 46.

seemed to deny the feasibility of careful, incremental change and improvement to the status quo; it was re-engineer or bust! Fortunately, as a management fad, its time was short. It made its contribution to the development of management theory, and the damage was reasonably contained.

In fact it seems likely that the 'productivity paradox' was a bit of a misunderstanding. Although there certainly was some unwise IT investment, it is likely that a combination of measurement errors and overlooked time-lag effects led to an incorrect analysis of the situation. And, although as discussed in Section 8.5 there are great reservations about this, people made redundant in one skills area were replaced by other people with skills in other areas. Fewer production people were required to generate the same level of output as previously, but more people were required in support areas, such as systems design and maintenance, computer programming, administration and such like. So although output rose, the macro level of employment also rose in such a way as to keep aggregate productivity more or less constant. This made it seem as though investment in IT was achieving little or no payback in terms of aggregate productivity – output per man-hour.

David Noble, in discussion of the early years of numerical control based manufacturing, provides a different perspective on this process: 'While the manufacture of robots, for example, is "expected to create 3,000 to 5,000 jobs," the robots themselves "will replace up to 50,000 auto workers".'[10] Lots of people lose their old jobs, others get lots of newly created jobs – just not the same people.

8.3.3 Phase 3: Later period – internal integration (at any price)

Gradually, as machines and machine systems became more intelligent – that is more computer controlled – it became apparent that significant further efficiencies could be achieved by linking them together, so as to enable information to pass directly from one to another. This principle applied to the deployment of computers themselves in manufacturing and also to their deployment in commercial and services industries.

There are a number of reasons for networking machines in this way: to integrate work across different processes, to orchestrate and control overall production processes, to monitor product or service quality, to provide status information to management at the local and overall levels, and simply to capture production data for use in planning and forecasting.

This technological approach developed in three phases:

- Materials requirements planning (MRP): Concerned mainly with inventory control, materials management, and purchasing and production scheduling.
- Manufacturing resource planning (MRP II): Integrates all a manufacturer's internal production related processes, from product design though procurement, materials management, production management, order

---

10 Noble, p. 348.

processing, and accounting. computer integrated manufacturing (CIM) is a variant on the MRP approach applied to the integrated management of information in engineering manufacturing organisations, where the central functions are the firm's CAD/CAM operations.
- Enterprise resource planning (ERP): Integrates everything. Builds on MRP II by adding human resource management, payroll and some external, supplier and customer links. So, as above, plus: logistics, distribution, shipping, invoicing, and related accounting functions.

The basic principles of the ERP approach are: agree on a single definition for every data entity encountered in the organisation's operations; identify the functional and logical relationships between all these data entities; and establish the 'ownership' of, and agree on update and view rights for every piece of data in the organisation.

ERP solutions are available in two broad types. The first takes the form of a single mammoth software package, usually implemented in modular fashion, based on the use of a single large database and a consistent user environment, with similar user interfaces, screen forms, reports and so on for all functional modules. The alternative, sometimes called an open, or 'best of breed' approach, involves similar data analysis and specification processes. However, individual business functions maintain their own separate, specialist applications and associated databases. The structure and content of each of these databases are made available to other applications so that the data they contain can be queried and shared in other ways without the need for data duplication or multiple data entry.

ERP is reported by many companies globally as having delivered substantial benefits. According to AMR Research, now part of Gartner Group, world-wide sales were $28bn in 2006 and sales were forecast to grow by 11 per cent annually over the succeeding five years.[11] (This was prior to the onset of the financial crisis.) But there have been some well-reported disaster stories, including some in the construction industry. Whether these are isolated cases, or whether they represent the tip of an iceberg whose underwater bulk is concealed by smart public relations and deft legal action, is difficult to determine. Certainly the single-database form of ERP seems to evoke strong reactions. Amongst the concerns are worries about the attributes of innovations listed by Rogers:

- the difficulty of seeing clearly the promised advantage over existing, often best of breed, solutions;
- the lack of compatibility with existing methods;
- the perceived complexity of ERP systems;
- the difficulty of trialling them
- the lack of observability.

---

11 Woodie, A., 'AMR Research Bullish on ERP Software Market', *Unix Guardian*, 2007, 4(7). http://www.itjungle.com/tug/tug072607-story10.html (retrieved 8 August 2010).

Further problems of perception include the 'bet the shop' level of commitment that is generally required for successful implementation; the sheer cost of implementation; mistrust between operations and the finance function, usually the ERP sponsor;[12] the formulaic – one size fits all – nature of ERP; and for strategic thinkers, the rigidity of such systems: spend years documenting how your company works today, force it into some standard process box, then pour cement over it.

Whatever the reality of individual ERP implementation projects, most commentators support strongly the basic objective of the ERP philosophy: a concerted effort to improve the quality of information used throughout the firm's processes. In theory, a system that integrates and consolidates all the significant data required to run the organisation should achieve this objective.

8.3.4 Phase 4: Today – external networking, supply chain management

MRP, MRP II, CIM and ERP are all primarily about the integration of processes and information, across different functions and locations, within the individual firm. Supply chain management (SCM) considers the firm in its business context and attempts to extend the idea of process integration and information sharing beyond the firm's boundaries, to include aspects of the processes and information of its customers and suppliers.

The term SCM came into use only in the 1980s, but in many industries, electronic forms of information exchange between firms and their customers and suppliers and even their competitors, have been commonplace for decades. Unlike ERP, SCM does not attempt to integrate all aspects of the activity of partners in the supply chain. It focuses on the interfaces between firms and on the passage of goods, services and information across those interfaces.

CNC and CAD/CAM, as described in Chapter 5, are the greatest examples of transformation brought about by improving the quality and flow of information within an individual company. SCM has transformed the quality and flow of information between companies to a similar extent. It is thus transforming not just the participating firms but also the entire industries in which they operate, as dramatically as CAD/CAM previously transformed the individual companies.

SCM is arguably the most important new management idea of the past 50 years. It has very quickly become clear that SCM depends more than anything else on the ability of trading partners to share accurate, up-to-date information with each other. The individual supply chain partners must therefore be able to generate the relevant data accurately and quickly. They must also be able to exchange that data accurately and quickly amongst themselves. 'Supply chain management

---

12 For an interesting, sceptical discussion of the role of the finance function in modern companies, see Pine, B.J., *Mass Customisation: The New Frontier in Business Competition.* Cambridge, MA: Harvard Business School Press, 1999, pp 121–128.

has become more about the management of information than the movement of goods.'[13]

8.3.5 Summary

The pattern of innovation diffusion outlined here: single-point systems, local optimisation, internal integration, and finally, external linkages has been observed across most of the sectors of the modern economy. This is essentially the pattern that has given rise to the transformation of those sectors in recent decades. The construction industry reader will be aware that, for the most part, at project level, this industry has hardly moved beyond the first of these four phases.

8.4 Major industries transformed by information technology

As suggested in the previous section, in terms of the functions they serve, modern information technologies have generally emerged in a progressive, evolutionary manner from identifiable earlier mechanical or electrical equivalents. And in many cases mechanical predecessor technologies persisted for many years after the general dissemination of digital systems. So it is difficult to draw a clear line, to say that the use of a particular technology started at a particular date or point in time. The following sections review the experience of major sectors in the US economy, focusing on the economic or operational conditions that drove the particular innovations, and the systems and associated standards that were implemented in response.

8.4.1 General manufacturing

The intense level of competition that manufacturers have faced, both locally and internationally, in the post-war period particularly, has forced businesses to reappraise fundamentally the nature of their production processes. All factors of production have been tested, but the emergence of low-cost labour centres overseas has put labour under particular competitive stress. Craft-based manufacturers and others with high labour inputs have had to make the greatest adjustments.

There have been three broad outcomes. Some domestic sectors, such as shipbuilding, clothing and footwear have been more or less eliminated in competition with low labour cost overseas producers. (Some manufacturers have survived as businesses, but only by relocating their production operations overseas.) A second group have survived by creating niche markets where craft working or other labour-intensive techniques are irreplaceable. The third group has survived by adopting modern, largely digital, technologies to dramatically improve their usage of both capital and labour. By the late 1950s, general manufacturers – producers of household goods, radios, TVs, bottled

---

13 Jacoby, D., *Guide to Supply Chain Management*. London: The Economist/Profile Books, 2009, p. 170.

goods, toiletries, paper and so on – were using early digital systems in the control of a range of production processes: feed and removal of work pieces, materials handling and packing and wrapping, for example. The main reasons for the deployment of these machines were, as ever: reduction in labour costs, elimination of human error, reduced scrap and waste, general reduction in inventory and significant reduction in manufacturing time.

8.4.2 Engineering manufacturing

Economic context / operational background

The late 1950s saw the introduction of the first numerically controlled (NC) machines in engineering manufacturing, particularly in the manufacture of complex products like aircraft, cars and major items of mechanical equipment. The origins and development of numerical control and computer numerical control (CNC) were outlined in Chapter 6.

The dissemination of CNC speeded up dramatically as the computers at the heart of CNC became increasingly affordable, with the introduction of minicomputers in the 1960s and embedded microprocessor chips in the early 1970s. The process was further accelerated by the release of early commercial computer-aided design (CAD) systems in the 1970s, and the subsequent ability to integrate CAD data directly into CNC systems, to produce what is now referred to as computer-aided manufacturing (CAM). These systems comprise a range of milling and turning machines used to create individual parts. The parts may be installed directly into the final product, or may first be combined into multi-part components or assemblies and then installed into the product. In most modern factories, most of the operations involved in these production steps are carried out using networked machines of various types, including industrial robots.

Manufacturing processes and their supporting systems became increasingly integrated during the 1970s and 1980s. Innovative approaches to manufacturing evolved from basic CAM, through techniques like flexible manufacturing systems (FMS), which enabled firms to re-program machines and production lines rapidly and easily, thus to 'mass' produce different products in relatively small batches.[14] The focus of management attention tended to move away from the technologies as such and more towards the analysis of their production processes. This led to 'just-in-time', 'lean manufacturing' and more general adoption of supply chain management approaches, all of which depend heavily on extensive, accurate information sharing between companies. By the mid-1990s scholars could write about manufacturing that: 'The impact of new technology ... has redefined expectations for quality, precision and the overall efficiency of the production process.'[15]

---

14 See, for example, Pine, *op. cit.*
15 Cohen. M.A. and Apte, U.M., *Manufacturing Automation*. Chicago, IL: Irwin, 2009, p.134 quoted in Cortada, 2004, p. 159.

(In addition to the general economic reasons for implementing manufacturing technologies as noted above, there were other, more politically motivated considerations. NC initially and later CNC, more or less eliminated the need for unionised, blue-collar, skilled machinists. These systems made production a function that could be run by 'trusted' non-union, white-collar workers in the engineering office. This was an important consideration at a time of labour unrest in the USA, the aftermath of the passing of the Taft–Hartley Act of 1947, of the odious Senator McCarthy, the Red Scare and all that. The publishing industry in the UK was another example where information technologies were introduced, at least in part, to reduce the influence of hitherto powerfully unionised workforces.)[16]

Technical solutions/standards issues

Business communications between supply chain partners nowadays typically involve relatively large flows of relatively small, relatively simple packets of commercial and operational data. These information flows are managed precisely and reliably through a combination of neutral data exchange standards and industry-specific, or even supply-chain specific, communications protocols, negotiated amongst themselves by the members of industry networks.

In an engineering manufacturing environment, two broad types of data must be shared:

- technical information, such as model data, CAD drawings, analysis data; and
- commercial information, such as material requisitions, purchase orders, invoices, production schedules and such like.

Communications between different computer systems in different companies is a two-part process in which data-sharing firms must agree:

- First, who provides what type of information to whom, by what means, at what point in their respective business or production processes. These are the communications protocols.
- The second area of agreement is on the precise structure and content of individual standard messages and the precise technical format in which the information is to be exchanged. These are the data exchange standards.

The standardisation of business data exchanges began in the early 1980s with the publication by the American National Standards Institute (ANSI) of the X12 standard, and internationally by UN/EDIFACT. X12 and EDIFACT are nothing more than very precise specifications of a wide range of standard business documents such as purchase orders, goods received notices, invoices and such like. So they describe the content and precise format of business messages. They say nothing about the communications medium or the communications protocols

---

16 Noble, p.238.

to be used. So X12 and EDIFACT messages can be transferred by dial-up modem, by e-mail, or by FTP for example. Value added networks (VANs), operated by firms like GE and IBM, dominated the early message-handling industry. As the internet has grown, web-based services have come to the fore, with messages, sometimes embedded in XML, being transported using secure web protocols such as AS2 and HTTPS.

The need for standards to govern the exchange of technical data was recognised and acted upon very early. The first version of the most important standard for CAD data, the Initial Graphical Exchange Specification (IGES), was published in 1980. Although it was superseded by an International Standard, ISO 10303 STEP (Standard for Exchange of Product Information) in 1994, IGES remains in use and is probably still the most widely used mechanical CAD/CAM standard, used both for data exchange and for data archive purposes.

In theory, IGES and its successors in STEP should enable all of the necessary engineering information, generated at the design stages of a new product for example, to be exchanged seamlessly between all of the different specialist applications in a given supply chain. In fact, the large manufacturers, such as car assemblers, tend to dictate to their supply chains how they want 'their' data delivered. This applies particularly to CAD/CAM data. In order to avoid the inevitable problems of translation and data loss associated with neutral exchange files, the assemblers prefer to mandate the CAD/CAM software to be used by their suppliers in designing the components that they manufacture. By adopting this approach they commit their suppliers to their business and eliminate data exchange or translation problems; less than ideal perhaps, but usually it works.

8.4.3 Process industries: petroleum, chemicals, pharmaceuticals

Economic context / operational background

The main purpose of digital technologies in these industries was to enable remote items of equipment and instrumentation to be managed centrally. They made extensive use of mechanical, electrical, hydraulic and pneumatic closed-loop control systems during the early part of the twentieth century. These devices had to be interrogated by eye and operated by hand, which was inefficient when plant was widely distributed and potentially dangerous in hostile production environments.

However, the experience of designing and operating these manual systems meant that when the first digital technologies became available in the 1950s and 1960s, they were taken up quite readily. The logic of closed-loop control systems could be emulated and automated accurately. Compared with manufacturing, process industries traditionally employed fewer, more highly skilled operatives in a less antagonistic industrial relations environment. As a result the owners' principal incentive was to optimise their production processes rather than lowering labour costs. This meant that, although there still was a significant reduction in employment in these industries over the years, the most remarkable developments were in the improvements achieved in their methods of production and distribution.

Technical solutions/standards issues

The process industries all used information technologies intensively to control the continuous production processes in their manufacturing plants, refineries and so on. The oil and gas industry also invested heavily in computer systems for other purposes such as seismic and other types of geophysical modelling of prospective producer fields. The industry operates extensive, shared, product distribution pipeline networks. So a third big area of technology investment was in the computers and embedded microprocessors used for optimisation, scheduling and accounting functions related to these facilities.[17]

Two classes of data are essential to these industries. The first is the data generated in the control of their production and distribution processes, generally referred to as supervisory control and data acquisition (SCADA) systems. As with CAD and other technologies, where there are many participating vendors and manufacturers, the SCADA community has developed a range of supposedly 'open' data standards that, in theory, provide for interoperability between equipment items.

The second class of data, as with manufacturing industries generally, is EDI documents and communications protocols. These industries embraced supply chain management seriously in the 1990s and have become major users of third party e-commerce services.

8.4.4 Retail

The retail sector's transformative use of IT has many lessons to offer the construction industry. Like construction today, retail started out on its programme of innovation as a complex, low profit, fragmented industry, with few apparent barriers to entry.

Economic context/operational background

Retail is the industry which has been most obviously, at least most publicly, impacted by information technology; transformed by electronic point of sale (EPOS) systems, the universal product code (UPC) barcode system, combined with EDI over value added networks and subsequently the internet. The change process was started in the early 1970s by a combination of manufacturers and retailers in the US grocery sector, who were under enormous pressure from a variety of forces. As described in the introduction to Stephen Brown's book,[18] the circumstances of the time were peculiarly influential:

---

17 Cortada (2004), p. 168ff.
18 Dunlop, J.T., and Rivkin, J.W., 'Introduction', in Brown, pp. 20–25.

- It was a period of high price inflation in foodstuffs worldwide. This was coupled with price controls in the USA, which resulted in retail food margins of less than 1 per cent. Firms were faced with a desperate need to reduce costs.
- Labour accounted for 67 per cent of retail expenses, apart from the cost of goods sold. And non-supervisory staff made up 90 per cent of the retail workforce, so firms placed an acute focus on labour-saving opportunities.
- Technology was crucial: the increasing availability of affordable, low-power lasers; optical character recognition software; powerful, smallish (mini-) computers capable of interactive, on-line transaction processing; and value added networks to connect stores in clusters or to head offices.

The main incentive initially was to reduce labour costs. But close second was the need to manage inventory more effectively. A large store holding 100,000 or more stock-keeping units (SKUs) might spend millions of worker hours a year in counting and ordering inventory.[19] A more widespread but less acute problem revolved around the poor quality of the management information used to run their companies. Because individual departments all used different data, and data capture techniques were inaccurate, companies were finding that reporting was poor, untimely, and provided little effective support to management decision making.

Technical solutions/standards issues

However, they found that: 'General purpose computers, by themselves, did not provide economically compelling industry-altering applications.'[20] As with other industries, only industry-specific technologies had the real power to transform; for retail that was the UPC/EPOS/EDI combination. The story of how this came about is a fascinating one, described in great detail in Brown's book. There are a few highlights of relevance to construction.

The unique product code concept originated in the late 1960s. The idea of the UPC is that every item a manufacturer produces should carry a label with a code on it that uniquely identifies the item in question. The label must be capable of being machine read, easily, accurately and quickly, even in wet, dirty or low-light conditions.

An extraordinary committee, comprising the chief executive officers of the leading food manufacturers and groceries and chain stores, was set up in 1970 to address the issue. In addition to the basic technical questions, the group was required to determine: 'whether such a code would actually be worthwhile, what it should be, how could widespread acceptance be achieved, how should the code be administered and should there be a standard symbol representing the code, and if so, what should it be?'[21]

---

19 Brown, p. 293.
20 Cortada (2004), p. 206.
21 Brown, p. 43.

The grouping which was called the Ad Hoc Committee on a Uniform Grocery Product Code, included: the president of H.J. Heinz, the CEOs of General Foods and General Mills, the chairman of Bristol Myers, on the producer side; and representing the retailers, the vice-chairman of the Kroger supermarket company, the president of Fairmont Foods, the president of the A&P supermarket company, as well as others of similar stature representing cooperatives, health and beauty manufacturers and so on, covering the entire gamut of grocery manufacturing and retail in the USA. It is a measure of the stress that this industry was under that such a gathering of people, whose relationships were fundamentally competitive and inherently confrontational, could have been assembled in the first place. It is a remarkable tribute to its members that it became the astonishing success that it ultimately proved to be. Stephen Brown acted as the legal counsel to the committee for the duration of its work.

There are three key aspects to the UPC code: how it should be structured, how it should be represented symbolically, and how it should actually be applied to products. The first two problems were dealt with relatively easily. It was agreed that the code should be twelve digits long. Each manufacturer would be assigned its own unique identification number or numbers; it was then given a range of sub-numbers that it could apply to its individual products. This is essentially the unique product code structure that applies today. The second big step forward was agreement on the linear barcode as a machine readable version of the UPC.

A third problem was to get agreement as to who should actually apply the code to individual product items. This task was going to be expensive and, as code marking seemed at the outset to benefit grocers more than manufacturers, the manufacturers were reluctant to agree to labelling at source. However, coincidentally with the Ad Hoc's Committee's deliberations, the US Food and Drug Administration brought in new rules about the labelling of foodstuffs, where these claimed to offer particular health benefits. This applied to a large proportion of the foods of the day, so a large proportion of labels would have to be reprinted to meet the new regulations in any case. The marginal cost of relabelling the unaffected items was felt to be tolerable to the manufacturers, so the project proceeded on the basis that manufacturers applied the codes.

With UPC and electronic point of sale (EPOS) systems, check-out tills became data capture devices, as well as cash collectors, with the result that over-stocking and stock-outs dropped dramatically, so sales rose. And, because the numbers of floor staff could be greatly reduced, staffing costs fell sharply, providing a double boost to increasing profits. These systems also helped increase speed and reduce errors at the checkout, which helped improve customer relations and increase customer loyalty.

The payoff for the manufacturers came in the form of huge improvements in their warehousing and distribution processes. Once a product item has had its barcode applied, every step in its journey to the checkout is recorded and can be managed precisely. An item can be checked off the production line, onto a truck, into a warehouse, onto a storage rack; off the rack, onto a truck, into a supermarket loading bay, into the back store, onto the shelf and through the checkout. All of the firms along this route – manufacturer, logistics firm, retailer – know exactly the current status of the item.

With a few minor rotations of individuals, the Ad Hoc Committee remained intact from its inaugural meeting in August 1970 until it 'faded away' in 1975. In June 1974, a packet of Wrigley's chewing gum became the first item carrying a final version of the bar code to be scanned through a checkout.

The Universal Code Council (UCC) was set up by the Ad Hoc Committee to manage and administer the allocation of individual manufacturer codes. The numbers of manufacturers' registrations grew rapidly, from a few thousand in the early years to over 110,000 in 1994 and 300,000 by 2006, not just in the grocery sector, but throughout retail and a huge range of other industries throughout the world. Today, 'the … system, is used by more than one million companies doing business in 150 countries across more than 20 industries'.[22]

One very strong message about the practicalities of standards-setting comes out of the Ad Hoc's workings. 'Pressure to move expeditiously was paired with equal or greater insistence to do the job well. In selecting a code, the Ad Hoc Committee had deliberately avoided the elaborate procedures of the American National Standards Institute. They feared, correctly, that the ANSI process, whilst exquisitely fair, would so elongate their task that success would almost certainly be foreclosed.'[23]

A second lesson to take from the retail experience relates to the comparatively slow and cumbersome process undergone by the industry in agreeing on the technical standards for electronic data interchange (EDI) between companies. Work started on this in 1974, and for a number of years the grocery industry operated its own version of EDI. For a variety of reasons, the chief executives delegated the development of the necessary standards to technical people; decision making probably suffered as a result.

The impact of UPC/EPOS and EDI has been truly transformational. A review carried out for GS1 in 1999 reported:

> Twenty-five years after its initial use, the actual impact of the U.P.C. on the nation's food industry was nearly 20 times greater than the original forecasts. Without the economic impact of the U.P.C., food prices to the consumer would have risen almost twice as fast over the 25 years. The conservative, initial estimate, originally forecast by the ad hoc committee, put savings to the food industry at $1.43 billion. The net benefits realized are 5.65 percent of sales within the grocery channel — with $293 billion in scanned value or $17 billion annually as of 1999. This is 50 times greater than the original estimate.[24]

---

22 http://www.gs1us.org/about_us/history, (retrieved 10 September 2010).
23 Brown, p. 60.
24 http://www.gs1us.org/about_us/history/pricewaterhouse_coopers_research, (retrieved 10 August 2010).

Two particular features of UPC were essential in achieving these benefits. First, UPC provided an industry-wide *lingua franca*, a common language that everyone in the supply chain could read, write and understand. Everyone described the products they dealt with in exactly the same way. There was no need for human intervention or interpretation in communications between links in the supply chain. Once a piece of product information was captured, its content and data about it, such as the location, date and time of capture, could be recorded and analysed by anyone in the community. Adjacent machines in the chain could pass on information in a smooth, efficient relay.

The second key feature of UPC is that it operates at the level of the individual product item. At this level of detail, there is no need and no scope for estimates or guesswork; things are binary: black or white; there, or not there. The data is irrefutable, and completely trustworthy. That means that in the communication between their machines and to the extent that they need to, to make the process work, firms can trust each other fully. The whole phenomenon of supply chain management is based on this concept of trustworthy information.

When the information in the supply chain is as accurate as this, it becomes possible to implement strategies like lean production, where accurate information on flows of materials and product enables the elimination of waste. Very high quality information is also necessary to implement agile manufacturing strategies where marketing, design and production departments in the participating firms all have access to shared databases of sales, logistics and production information.

A particularly interesting result of the implementation of these technologies in retail was the change in relationships between customer, retailer and manufacturer that it brought about. Previously, manufacturers used to produce what they thought their end customers wanted, on the basis of information like fashion magazine reports and general trends, as they saw them. The retailers' buyers simply took a guess at how many of a particular item on a manufacturer's list he or she though would sell and ordered accordingly, often with disappointing consequences.

In the new supply chain, the retailer takes the dominant role. By capturing the purchasing preferences of its customers at the checkout counter, the retailer builds up a very detailed and dynamic picture of what is selling today, what needs to be replenished, and by how much. A famous example of how this capability has intensified is in the arrangement established between Wal-Mart and Procter & Gamble in the late 1980s, whereby checkout data is passed from Wal-Mart stores directly to P&G production systems as part of a continuous product replenishment processes.

These information technologies have enabled retailing giants like Wal-Mart in the USA and Tesco in the UK to harvest truly vast quantities of data on their customers' buying habits. This gives them very effective control over their supply chain partners and has led to a marked increase in economic concentration in parts of the retail sector.[25] This may have been a temporary phenomenon. Consolidation certainly seems to have occurred during the initial, high-cost, EDI era. However

---

25 Brown, pp. 14–15.

the degree to which internet trading has enabled anyone, with an idea and the ability to set up a website, to create entirely new businesses, has probably rolled back that initial consolidating impetus somewhat.

8.4.5 Financial services

Economic context/operational background

The post-war period was a golden era for the American economy; industrial activity accelerated rapidly, wages rose, standards of living rose and the demand for financial services of all sorts soared. Throughout the 1950s and 1960s and into the 1970s, the financial services industry – banking, insurance and broking – was characterised by low levels of competition, high fees, strict regulation, and significant fragmentation.[26] This was a flabby, comfortable industry whose members had little impetus to change and no incentive to take risks in doing so.

At least initially, the financial services sector failed to respond to the historic rate of growth of the American economy. The industry appeared to be either incompetent or corrupt, or both. 'Outside the industry … the exchanges and brokerage firms were all treated as one dysfunctional sector of the economy … (The Securities and Exchanges Commission [SEC] was deeply critical,) … poor practices, discrepancies of records, misuse of funds, and even theft. By the late 1960s, theft alone had grown to over $100 million in lost securities.'[27]

The main problem was that the accelerating demand for financial products brought with it ballooning volumes of the paperwork inevitably generated in a transactions-based, highly regulated environment. Each of the three sectors experienced its own particular pinch point: in banking, this was cheque clearing; in insurance, it was policy calculation and issuance; and in broking it was in handling enormous numbers of fast-moving buy/sell transactions. The problem of setting up and managing huge numbers of customer accounts was common to all three.

Technical solutions/standards issues

Two things happened. First, simply to survive the avalanche of paperwork, firms in all three sectors had to throw enormous sums of money and huge IT at this problem. This was not with the intention of changing or improving their processes, but simply of surviving by making them faster and cheaper.

In banking for example, a number of industry committees and organisations had already agreed on basic but critical issues such as the size and layout of a standard cheque. They also agreed on magnetic ink and optical character recognition standards. The industry was also a very early user of inter-firm networks, so all in all, quite a lot of ground work had been done by the early 1960s. Nevertheless, the problem of inter-bank transfers of cheques was still huge. A 1964 report calculated that cheque clearing in the USA involved 150

---

26 Cortada (2006), p. 29.
27 Cortada (2006), p.165

billion separate cheque handlings annually.[28] The banks used electronic funds transfer systems (EFTS) extensively and the introduction of ATMs eased the load somewhat, but these services complemented cheque issuance, they did not substitute for it. The number of cheques issued seemed simply to parallel the path of economic growth.

Gradually however, bankers began to use IT in more strategic ways, as Cortada reports: 'By the end of the 1980s bankers were using computing to sell more or different services not merely to hold down costs.'[29] And: 'By the mid-1990s CEOs were in office who had personally experienced implementing various IT applications. For them, IT had become strategic, so they were involved.'[30] During this period a wide range of new products and services were developed, the most notorious being the collateralised debt obligations (CDOs) which were largely responsible for the 2008 Financial Crisis. More recently, particularly since the establishment of the first internet bank in 1995, the internet has brought about some significant changes in banking, with web-based account management and non-bank payment services such as PayPal.

Insurance companies used computers for managing customer accounts and for their actuarial, underwriting and investment management activities. The first of these is relatively simple but hugely labour-intensive work, and was early and easily moved on to computers. The large amounts of paperwork associated with the other activities were also computerised at an early stage.[31] The higher level work continued to be done by humans, although even here computers were increasingly being introduced.[32] And as with banking, the insurance industry has developed a range of internet services including marketing, purchase of policies and account management.

The same general pattern applied in the case of stock and bond brokers and exchanges. The paper problem was very acute for brokers. In 1968, brokers started closing their offices at 2pm rather than 3.30pm in order to clear the day's transactions. In the late 1960s there were 700 brokerage firms on the New York Stock Exchange (NYSE); by the end of the 1970s 'the paper crisis had killed off 200 of them'.[33]

The second important initiative came from governments and regulators, who saw the incompetence of the broking industry as a real threat to economic growth and applied huge pressure to them in an effort to bring about significant improvement. Many of the industry associations responded; special study groups and task forces were set up. But this was in the face of dogged objection and antagonism from individuals and firms who resisted the adoption of computing 'because they knew how it could harm them personally'.[34]

---

28 Cortada (2006), p.44.
29 Cortada (2006), p. 93
30 Cortada (2006), p. 100.
31 Cortada (2006), p. 115.
32 Cortada (2006), pp. 121, 129.
33 Cortada (2006), p. 165.
34 Cortada (2006), p.187.

Ultimately of course the regulators won. NASDAQ started operations in February 1971. Fixed fees were banned and the beginnings of a nationwide market were established in May 1975. Paperless trading started in 1995; 80 per cent of trades were paperless by 2007. London had its de-regulatory Big Bang on 27 October 1986. The Y2K scare brought about the replacement of virtually the entire inventory of IT systems, both software and hardware.

To get some idea of the scale of investment in the brokerage industry over the last two decades of the twentieth century, consider the fact that in the whole of 1950, the NYSE handled about 525 million transactions. In 1990 that number was 40,000 million; by 2004 it had grown to 367,000 million transactions per year.[35] In what was called the 'Flash Crash' of 6 May 2010, against a background rate of about 10,000 transactions a minute, the actual crash was caused by a spike rate of 80,000 transactions per minute.[36] By the end of the 1990s brokerage was the most computerised of all the financial services industries – which made it one of the most computerised sectors in the entire economy.

Brokers were also early adopters of the internet. Almost all organisations launched marketing sites in the mid-1990s. By 1999, 16 per cent of all transactions were online and by 2000 a third of all retail transactions were web based. The industry continues to offer an increasing range of data and transaction services, particularly for retail customers. Online brokerage was 'fundamentally changing the relationship between the broker / dealers and their customers'.[37]

8.4.6 Other industries

Information technology

In a boot-strap sort of way, the IT industry itself as it exists today would be impossible without some of the most powerful products of the IT industry. Chip manufacturers depend completely on CAD/CAM and computer-integrated manufacturing techniques to achieve the tolerances necessary to continuously drive down the size and increase the transistor density of integrated circuit boards. Improved manufacturing productivity and quality derive directly from the industry's enhanced use of computing.[38] And the hard disc drive industry, arguably the most complex of all areas of electronics, simply could not exist without the advanced supply chain management systems and techniques that they have used for most of the past two decades. Although in many respects software programming remains a craft – labour intensive and reliant on extensive, rather traditional skills – even this is improving, as higher level languages, code reusability and better, computer-enhanced design techniques are introduced.

---

35 Cortada (2006), p. 156.
36 'The Flash Crash Autopsy', *The Economist*, 9 October 2010, pp. 95–6.
37 Cortada (2006), pp. 183–4
38 Cortada (2004), p. 285.

158 *IT usage in construction and other industries*

Transport

Until recently, it would have seemed unlikely to suggest that the main freight transportation industries – trains and trucking – might be serious consumers of IT. Yet today, as Cortada puts it: 'The management of information is becoming as important as the management of freight for the trucking industry.'[39] Not really such a surprise perhaps, when one considers transport in the supply chain context, and the key trend of integrating transport into supply chain systems so that manufacturing, distribution and retail sectors could share data, '… the glue that held compatibility and integration together from one sector of the economy to another …'.[40] Computers, which were used in the 1950s, 1960s and 1970s in these industries primarily to reduce labour costs and to optimise routing and load factors, had by the end of the 1990s become an essential part of the global supply chain. 'As in the manufacturing industry, in the transportation industries we have seen a shift of knowledge, cognitive behaviour and responsibility away from workers towards computers.'[41]

Cortada goes on in the rest of volume two of *The Digital Hand* to discuss a whole range of other industries: telecommunications; book, newspaper and magazine publishing; radio, TV, movies and recorded music, video games and photography. Volume three considers the public sector: tax, defence, law enforcement, social security, postal service and educational systems. In every case, digital technologies have played an important part in creating organisations and institutions that are fundamentally different in structure and mode of operation to their equivalents of 50 years ago. Hardly an area of the economy remains that has not been transformed in the sense described here by computers and communications technologies.

8.4.7 *General observations*

Over the past 60 years the digital revolution – the introduction of a wide variety of information and communications technologies – has brought about transformative change in almost every sector in the modern economy. The different sectors varied significantly, in terms of the impetus that drove the adoption of new technologies, the particular technologies adopted, and the consequences of the initial adoption actions. However they all had one thing in common: none of the firms or industry sectors in question embarked on their adoption journeys willingly or enthusiastically. As Cortada points out time after time, in virtually every case firms invested in new technologies only under irresistible pressures of one sort or another.

Thus, aircraft manufacturers were compelled by the US Air Force to adopt CNC technology, because the USAF was convinced that safe jet aircraft simply could not be manufactured using conventional engineering techniques. General manufacturers adopted supply chain type techniques using EDI and X21 at least

---

39 Cortada (2004), p.250.
40 Cortada (2004), p.228.
41 Cortada (2004), p.257.

partly in response to the threat of their business emigrating to low labour-cost manufacturers overseas. Grocers invented UPC and EPOS as a desperate attempt to rescue profits which were being devastated by high input costs and price controls. Banks and brokerage businesses were forced to adopt new technologies in order to survive the tidal wave of paper generated by the long post-war economic boom. Regulators forced them to compete by eliminating standard fee scales and restrictive practices. They all invested heavily in EDI and other business-to-business communications capabilities.

Each sector started from a position of resistance to innovation. Each eventually, reluctantly, took the innovative step and in doing so each sector unleashed its own particular technological genie. The individual sectors of the US economy are intensely competitive environments, so that as soon as the innovators in a sector started to see benefits or at least started to boast of such benefits, everyone else had to follow.

CNC and CAD/CAM were transformational at the level of the individual firm – they changed fundamentally the way in which the firm went about its work. Other transformations generally came about as a result of companies collaborating at supply-chain level. In all cases a crucial part of the innovative process involved the development of important technical standards and their adoption by their industries. Each of the main standards emerged in one of four main ways: through the work of formal committees of technology experts (BSI, ANSI, ISO etc.); as a result of its being mandated by dominant companies in an industry (e.g. CATIA in aerospace); as a result of the success of a particular technology vendor in dominating its market (MS Word, AutoCad etc.); and most unusually, through the work of committees of non-technical business leaders (UPC/barcode/EPOS by grocers). It is not difficult to guess which type of process gives rise to the best results, in the shortest time, at the least cost.

As described earlier, the process of industry transformation in most cases was a progressive one. The general pattern was for firms first to automate individual functions, then to link the systems controlling operationally adjacent functions. Next they integrated groups of functions across the firm, and finally they linked their internal operations and systems with those of their neighbours in their respective supply chains. The aim of each of these steps was to reduce costs, to improve efficiency, through inventory and process management and resource optimisation, and ultimately to improve management control. With a few exceptions, these moves were generally undertaken tactically, step by step, not as part of any greater strategic plan. In most cases they took at least 10–15 years – a managerial generation – to play out fully.

It is important to note that each of the steps outlined here required that IT systems associated with individual production units should be capable of producing well-specified, accurate data. As the integration processes developed, it became increasingly necessary for these systems to be able to communicate with each other. This in turn required that the data exchanged between systems was presented in highly standardised forms and that the processes of data interchange conformed to strictly agreed protocols. So the general pattern was: accurate, structured data to begin with; well-specified data exchange formats; and strict interchange protocols, detailed agreement as to who provides what information, to whom, at what point in the process. This is essentially the same challenge as the one confronting the construction industry in its implementation of the BIM approach.

8.5 Social consequences

8.5.1 The information revolution

It would be nice to report that the transformations described here and in James Cortada's books have been universally beneficial, that everyone whose life has been touched by them has been enriched. This is patently not the case. The competitive struggle in the modern economy can be as brutal as any in nature. For the consumer, the results are higher-quality goods and services at lower prices. But for the producer, it demands continuous refinement of his products and adaptation in his methods of production. Anything that improves his total factor productivity (output per dollar's worth of labour and capital) will be seized upon. Any overlooked opportunity could be fatal. For the individual worker, price competition in the digital economy particularly can be devastating.

Consider earlier industrial eras when fundamental technologies changed: water wheel, steam engine, internal combustion engine, electricity, factories, production lines. These all involved processes of transition, periods in which the labour force could re-train, learn the necessary new skills and gain employment in the new environment. The information required in the production process is embodied in the skills of the labour force.

The technology of the present era is information technology. We tend to think of IT as being mainly about computation and communications – which of course it is. And, as such, IT affects our economies and our lives very directly. But by far the most profound property of IT is the fact that this essentially is technology that embodies information. Anything that can be programmed will be programmed, if doing so increases productivity or reduces costs in some other way. And, while the relative cost of IT continues to plummet, it is to be expected that more and more human skills and technical information will become embodied in software, embedded in systems.

This is not like the transition from the putting out system of production to manufacturing in factories, or from early craft-based factories to production lines, from steam to electricity, or any early changes of industrial style. In previous industrial transitions there has always been scope for individual managers, supervisors and workers to carry their skills across from one mode of production to the next. It was not always easy or direct, as labour revolts from the Luddites onwards showed. But it remained the case that although the motive forces and production contexts changed, the basic nature of their work didn't change – the information they used remained fairly constant. The information they worked with was largely in their heads and in their hands. So the individual manager, supervisor, or worker was a necessary and valuable resource, both as a source of labour and as a container of information. Early forms of mechanisation dispensed with the need for humans as a source of labour, but they actually increased demand for humans as containers of information. Information technology may dispense with humans as information containers.

Opinions differ on this subject. Early in the history of the digital era, Herbert Simon, one of the pre-eminent American philosophers and social scientists of the twentieth century, described manufacturing industry of the future as comprising:

> An underlying system of physical production and distribution processes, a layer of programmed (and probably largely automated) decision processes for governing the routine day-to-day operation of the physical system, and a layer of non-programmed decision processes (carried out in a man-machine system) for monitoring the first-level processes, redesigning them, and changing parameter values.[42]

Earlier still, Peter Drucker, arguably the leading writer on business management, wrote: 'The popular belief that the new technology will replace human labour by robots is utterly false.' His argument was twofold: first, that the increased productivity brought about by automation meant not that fewer workers would be required to produce the same amount of output, but that more output would be produced by the same number of workers. He also went on to say: 'Actually the new technology (though there will be problems of displacement) will employ more people and above all, more people who are highly skilled and highly trained.'[43]

8.5.2 Polarised societies

A less optimistic picture is painted by David Noble. For him the 'problems of displacement' cannot be concealed in protective parentheses. He lays them wide open in a devastating critique of the whole programme of technological progress. For a construction industry that has yet to start down that road, this is a disconcerting work; the epilogue is truly chilling.[44] A readable, if shrill exposition of many of the same arguments is provided by Jeremy Rifkin.[45]

The theme is carried forward in a recent paper by a leading contemporary labour economist, David Autor of MIT, in which he describes the 'polarisation' of American economic society into relatively high-skill, high-wage jobs and low-skill, low-wage jobs as a result of the application of information technologies. In

---

42 Simon, H.A., *The Shape of Automation for Men and Management* New York: Harper Torchbooks, 1965, p.110, quoted in Cortada (2004), p. 380.
43 Drucker, P.F., *The Practice of Management*. London: Heinemann, 1955, pp. 34–5.
44 Noble, D.F., *Forces of Production: A Social History of Industrial Automation*. New York: Oxford University Press, 1984, pp. 324–353.
45 Rifkin, J., *The End of Work*. New York: Putnam, 1995.

proposing that something like the Simon model may have arrived already, Autor divides work into three classes: routine, abstract and manual.

> Routine tasks are middle-skilled cognitive and production activities such as bookkeeping, clerical work, and repetitive production tasks. The core job tasks of these occupations in many cases follow precise, well-understood procedures. Consequently, as computer and communication technologies improve in quality and decline in price, these routine tasks are increasingly codified in computer software and performed by machines.
> Non-routine tasks can be roughly subdivided into two major categories: abstract tasks and manual tasks. These tasks lie at opposite ends of the occupational-skill distribution. Abstract tasks require problem solving, intuition, and persuasion. Workers who are most adept in these tasks typically have high levels of education and analytical capability.
> Manual tasks, by contrast, require situational adaptability, visual and language recognition, and in-person interactions. Examples of workers engaged in these tasks include janitors and cleaners, home health aides, construction labourers, security personnel, and motor vehicle operators. Manual tasks demand workers who are physically adept and, in some cases, able to communicate fluently in spoken language.[46]

It is clear from Autor's work, that of Michaels *et al.* and others, that this phenomenon is not restricted to the USA, but is already widespread throughout the industrialised economies.[47] Jobs at the centre of the economy – those which comprise mainly routine tasks – are increasingly being computerised, or offshored to cheaper overseas locations. (Offshoring and computerising are closely analogous processes in that they both require that the service or manufacturing process involved must be very precisely specified before either can be applied. In a sense, offshoring could be considered an inexpensive version of computerisation.)

The conventional economic response to these developments is to argue that both increasing mechanisation/computerisation and increasing offshoring result in cheaper goods and services in the domestic economy, and are therefore beneficial to the economy as a whole. Further, as overseas producer countries become wealthier as a result of their offshoring activities, they in turn become consumers, creating a middle class with increasing demand for the supposedly higher value goods and services produced in the home economies.

So supposedly, on aggregate, the domestic economy as a whole, its surviving producers, and its consumers are all made better off as a result of computerisation or offshoring. This fact will be of little comfort to the individual worker whose job

---

46 Autor, D., *The Polarization of Job Opportunities in the U.S. Labor Market: Implications for Employment and Earnings*. Washington: DC: Center for American Progress and The Hamilton Project, 2010, pp. 2, 4.
47 Michaels, G., Natraj, A. and Van Reenen, J. *Has ICT Polarised Skill Demand? Evidence from Eleven Countries over 25 Years*, Discussion Paper No 7898. London: Centre for Economic Policy Research, 2010.

is of the routine type and can therefore readily be offshored or computerised. And, though not much of construction is amenable to offshoring, there are many jobs in the industry that, relatively soon, will be mechanised or computerised, or will be eliminated simply because improved information in construction processes will render them unnecessary.

Autor's classification and the experience of other industries suggests that the sorts of construction jobs that will be eliminated by mechanical automation will be those of skilled trade supervisors and craftsmen, and clerical jobs of many types. Computerisation on the other hand will threaten people called knowledge workers. It is difficult to see how work in areas like basic drawing production, code compliance, technical analysis, scheduling, and quantity surveying might survive. Complete, accurate information generated in the course of design and construction will eliminate the need for large numbers of administrators, auditors and compliance officers, particularly in areas like safety, quality assurance and environmental management. Dramatically improved, streamlined procurement and supply chain management processes will require far fewer levels of commercial management and bureaucracy.

If progress in the diffusion of information technologies in construction were to continue along the pattern seen to date in the industry, this situation would be quite a long way in the future. However, as the case studies in Chapter 7 show, the basic BIM capability – intelligent parametric models, standard data exchange formats and agreed information protocols – is maturing quite rapidly in much of the industry.

It took quite a long time for people to become comfortable with the initial idea of designing with computers, using conventional drawing-based CAD systems. But now that that psychological barrier has been largely overcome, it seems reasonable to suggest that people are likely to take to designing with models rather more readily. A critical take-off point, where the building design community generally becomes comfortable with the use of BIM tools and protocols, can be expected to be reached within the next five years or so. When that happens, a period of dramatic, disruptive discontinuity in the industry is likely to follow. Abrupt changes of operational mode will take place particularly amongst those companies involved in larger construction projects, say those worth £10m or more. This part of the construction industry really will be transformed.

The impetus behind the change will come from three main sources. First, well-informed, 'intelligent' clients will see the opportunity to obtain their buildings on a competitive fixed price, lump sum basis, and will demand that service from the industry. Secondly, design firms will see the opportunity to regain ground and influence lost over the past 20 years or so and will encourage their clients to pursue this approach. But, probably of greatest importance, leading contractors will see the opportunity to create and own this new market and will quickly grasp the initiative. Only the most operationally competent, best-capitalised companies will succeed in this.

The guiding rule in the overall process will be: anything that is routine can be programmed and anything that can be programmed will be. Companies will invest heavily in capturing, in systems form, the information used in their business processes; systems are simply too cheap and too reliable for things to be otherwise. Most forms of knowledge, including most forms of tacit knowledge, will eventually be codified and embodied in systems of one sort or another. These systems will become the principal assets, of professional firms particularly. Only a small proportion of the people currently employed will be needed to animate them. The key qualities of these remaining people will be capabilities for which computers cannot be substituted: intuition, judgement, interpersonal communications skill and imagination.

8.5.3 The significance of decision making

Direct observation and other methods of recording events in the natural world give rise to the fundamental entities we call data. Information is created by the application of rules to specific sets of data. This imparts structure to the data, which makes it amenable to being harvested, organised and presented for use by people.

Every conscious human action is preceded by a decision. A decision results from the application of human judgement to a quantum of information. All other things being equal, the 'quality' of the decision is dependent on just two things: the quality of the information on which it is based and the quality of the judgement applied in the decision-making process. In this sense, information is the most important external input to human decision making and conscious activity.[48] In this sense also, knowledge is not a thing, it is a state of mind. It is the state of mind of a decision maker, on the point of making a decision; the state of mind necessary to effectively combine the information at his or her disposal with the judgement necessary to make the decision.

The most important internal, psychological, input to decision making is the attribute called judgement; a slippery combination of intuition, instinct, training, experience and, above all, imagination. The body of this book started, in Chapter 3, with a critique of the particularly dangerous combination of information and judgement on which the construction industry currently depends to get projects built. Almost every area of work associated with the design, procurement, construction and operation of buildings involves huge amounts of conscious, deliberate decision making. The drawing-based information currently available to support these decision processes is of fundamentally poor quality, as described in Chapter 3. This means that people must compensate by applying disproportionately high levels of judgement in the decision making required to carry out their work. Judgement is an erratic, human, attribute; when it fails, decisions go awry and projects fail.

---

48 For a more formal, very interesting discussion of decision making in construction see: Kam, C.K.H., 'Dynamic Decision Breakdown Structure: Ontology, Methodology and Framework for Information Management in Support of Decision-Enabling Tasks in the Building Industry', PhD. thesis, CIFE, Civil and Environmental Engineering Dept., Stanford University, 2005.

Information generated and communicated using BIM methods is of spectacularly better quality than traditional, drawing-derived information. The level of judgement required to use it well in decision making is commensurately lower. The result is that fewer decisions should go wrong, fewer projects should fail.

However, while BIM will improve the quality of information used in industry decision making, it will not significantly reduce the numbers of decisions that have to be made. And the animating judgement required in making these decisions will continue to be provided mainly by people, at least for the reasonably foreseeable future. So, for the foreseeable future, the construction industry will continue to depend on relatively large numbers of decision-making people. These people will be using very high quality information, but wielding more or less the same traditional powers of judgement and imagination; powers that machines have not yet acquired. Because of its 'decision density' – the relatively high requirement for decision making – construction may not suffer as much as other traditional industries have from the hollowing out of work described in the previous section.

8.5.4 Technological progress, productivity and employment

Nonetheless the level of employment of people of almost all sorts throughout the large projects industry particularly, will fall steeply in the medium term – say the next 20 years or so. These people make up quite a large proportion of the overall construction workforce, which means that they also represent a significant proportion of the total national economic workforce. The relatively small numbers left at work will, on average, be much more highly paid than previously. But small numbers of big spenders have far less economic impact than large numbers of moderate spenders. And, logically, in replacing people with systems, construction companies on aggregate will spend significantly less on wages, so the overall spending power of the industry workforce will diminish substantially; aggregate demand will fall correspondingly.

The conventional economic view of this is that, in the long run, increases in the aggregate level of productivity do not result in increases in the natural rate of unemployment in the economy.[49] The conventional argument holds that increasing productivity lowers prices, thus making the affected goods and services affordable to larger numbers of consumers, which means that real wages fall, causing demand for labour to rise. At the aggregate economic level, the forces of 'creative destruction'[50] that destroy one set of jobs, simultaneously create a whole new set of jobs in more productive, higher value added industries or elsewhere in the economy. Demand for bricklayers declines, but this is offset by increasing demand for hairdressers, computer programmers, and so on. People just need to be re-trained and re-deployed to cope

---

49 See for example, Blanchard, O. and Katz, L.F., 'What We Know and Do Not Know About the Natural Rate of Unemployment', *Journal of Economic Perspectives*, 1997, 11(1): 56.
50 Schumpeter, J. A., *Capitalism, Socialism and Democracy*. New York: Harper, 1975 original publication 1942.

with the transition. That at least was what was supposed to have what happened when UK coal mines, steelworks, shipyards, footwear and other old-fashioned manufacturing operations were shut down or exported to Korea.

The social devastation that actually resulted from these events was concentrated on individual, local communities. So although the experience of trauma was intensified, local social structures provided some degree of mutual support and remedial efforts could be focused fairly accurately on the affected groups. There would be no devastation of homogenous industry communities if the same sort of thing were to happen to construction, but equally it would not be possible to provide the sort of focused support that was directed at these older communities.

The impact of BIM on construction will be at least as profound and as fast acting as the experience of those other old industries. Two issues arise. First, will the Schumpeterian assumptions hold in the face of the sudden loss of such a large pool of labour and the concomitant reduction in spending power? Will alternative jobs really materialise elsewhere in the economy to compensate for the collapse in construction employment? And if they don't, how will the economy cope with the sudden loss of such a large element of aggregate demand? Is a natural rate of unemployment of 12–15 per cent conceivable or tolerable? Perhaps a BIM-induced transformation of construction might be a big enough event to tip the economy as a whole out of Schumpeterian dynamic equilibrium? Is this part of a larger, uglier process leading to something like Rifkin's 'end of work' scenario? Are we even perhaps witnessing the beginning of the self-destruction of capitalism, overwhelmed by its own internal contradictions, as forecast by Marx?

The second set of questions concerns the way in which individual construction people and their firms adjust. Will construction really be transformed and if so, will its transformation be just like that of these other large old industries? Perhaps BIM is just another of the fads, spasms, that periodically energise industry leaders and commentators? Or, with BIM, will construction embark on a new, modern, enlightened way of doing business. Could the whole way in which people think about construction – the industry paradigm – be changing for ever?

There are no true or complete answers to questions like these. But the purpose here, as elsewhere in the book, is to open up the issues – particularly the remarkable combination of threats and opportunities – that now confront the industry and those associated with it. Individual people, the organisations they belong to, the industry at large, and government, all have an interest in this discussion. We cannot hope to change the future, but by imagining what it might bring, we can at least prepare ourselves for its effects.

8.6 The pattern of IT adoption by construction firms

The construction industry spends nearly as much per head on IT as other industries, including those considered earlier in this chapter.[51] Whereas all of

---

51 The Knowledge Practice, Construct I.T. (Salford University), *Building on IT 2010*. Manchester: National Computing Centre (NCC), 2010.

these others have been transformed by their IT investments, construction has experienced no comparable change or improvement in its performance. It is important to understand why this has been the case. This section considers some possible causes.

The work of most companies in the construction industry can be broken down into three broad areas of activity:

- administration, including purchasing, accounts, human resources and payroll;
- job winning, largely the role of the estimating and engineering/technical services departments, business development and company directorate;
- project execution – what everyone else does.

Administration and accounts functions in construction are largely similar to administration and accounts in other industries, with reasonably well-established procedures and standards. Information management requirements tend to be relatively orthodox, so data structures and computing processes can be reasonably well specified, and conventional business applications can be used, more or less with confidence. The interfaces between corporate accounting and project material control and cost management systems can sometimes be less than ideal but in general, nowadays, they work and support the necessary audit trails and other financial verification procedures.

8.6.1 Job winning

The business development aspect of job winning is fairly generic in construction, so standard customer relationship management (CRM), and basic contacts management systems should work reasonably well. It is sometimes felt that client relationships, notably those with repeat clients, are particularly personal in construction. As a result, problems can arise where individual regions or divisions don't fully share 'their' client information with other divisions or regional offices.

The main job of business development is to win opportunities to bid for projects. Once an invitation to tender has been received, estimating and engineering/technical services take over. For particularly important tenders, key members of what will ultimately be the project delivery team might be pulled out of their current projects to help with the development of the tender. Their contribution will be largely presentational; they are unlikely to make a substantial contribution to the commercial components of the tender, but at least they will be aware of the project to come.

Cost planning and estimating – calculating the cost of a project as part of the contract tendering process – are both black arts. A number of software tools and pricing services are available to help in analytical estimating; building up an estimate from the details contained in a bill of quantities or simply from a set of drawings and specifications. Pricing information is also available for the higher level, top-down activity of cost planning; calculating elemental unit costs or gross cost per square metre.

But both of these activities are subject to error from a number of sources: poor design information, inaccurate quantity take off, incorrect and mis-applied unit rates, and lesser errors, like incorrect overages and allowances. These are just a few of the specific problems that are inherent in the processes. They are all information-based problems, and currently they are overcome only through the application of the most subtle judgement and keen intuition of experienced commercial managers. Incorrect decisions and choices in this area can have drastic consequences.

As noted, a variety of data providers and software packages are available to assist in the production of cost plans and estimates. However, the tools most widely used in the calculation and presentation of cost plans and estimates are spreadsheets, and home-made applications built around spreadsheets and user-level database packages. The key feature of these systems is their great flexibility. The user can define data fields as he or she requires and can build structures and relations between data entities to suit his or her particular purposes and his or her particular view of the data. The spreadsheet is a type of data manipulation tool that supports very neatly the combination of poor or questionable input data with the application of judgement and intuition. Skilful builders of spreadsheets can produce tabular and graphical results that look extremely impressive, concealing effectively the flaws and weaknesses in the underlying data. They can give to poor-quality data an appearance of plausibility that deters challenge and creates a false sense of confidence in the calculation processes.

A further problem with this approach is that the detailed scratch-sheet type calculations are not generally included in the finished spreadsheets. This means that low-level calculations such as build-ups cannot be replicated. In a sense, only the original creator of the spreadsheet knows and understands exactly what it contains and what it represents.

The engineering/technical services department will usually prepare the method statements and procurement and construction programmes included in the tender documentation. As with the cost plans and estimates, the accuracy of project programming is entirely dependent on the quality of the design information on which it is based. The quality of the programme will also depend on the quality of the company's historical information, such as man-hour usage rates used in calculating activity durations and their lead and lag relationships. Again, a special combination of judgement, intuition and experience is required to manipulate the available information so as to provide a credible proposal; one that appears achievable and that also appears to meet the client's schedule objectives.

This is generally done with planning and scheduling packages of various types. These share many of the weaknesses of spreadsheets, in terms of information quality. (See for example Peter Morris's critique of such systems: '… in many respects still stuck in a 1960s time warp'.)[52] Like a spreadsheet system, a good planning package in the hands of a skilled planner can make fundamentally

---

52 Morris, P.W.G., *The Management of Projects*. London: Thomas Telford, London, 1997.

IT usage in construction and other industries 169

questionable data appear much more plausible and well-specified than it really is. And, as with estimating systems, many of the detailed, build-up calculations are carried out outside the planning package itself and are not usually provided as part of the programme. So again, only the originator of the programme knows and understands exactly what it represents.

Spreadsheets and planning systems both enable users to take poor-quality data and, by applying their personal judgement and intuition, apparently transform it into plausible, trustworthy information. It is hard not to be impressed by some of the graphical output generated by skilled users of these tools. And the ability to crunch large volumes of data encourages them to create massive, complex models (spreadsheets and CPM networks) which tend further to deter challenge or meaningful questioning of the underlying inputs and the user's assumptions.

This is similar to the process by which drawing-based CAD systems can lead the user of the documents they produce to believe that the material they contain is higher quality than it really is: consistent, coordinated, clear, complete and correct – as discussed in Chapter 3. In fact, because of the unstructured, unsystematic way in which they handle data, all three of these application groups should really be treated with great caution. They are powerful tools, but used as they commonly are in construction, they can mislead managers and seriously impair the productivity of the project team. At the simplest level, because it is so easy to do so, they can be used to generate vast quantities of paper of dubious value, all of which must be managed and tracked as it flows around and clogs up the arteries of the project organisation.

The problems at the heart of this discussion originate in the poor quality of information generated by conventional drawing-based architectural design. This information forms the basis of all subsequent production and commercial aspects of construction project management (Table 8.1). Logically, these functions should all be based on the same underlying data and they should all tell more or less the same story about the project. But this almost never happens, or if it does it happens by accident.

At the detailed level, each of the disciplines involved tends to have its own view of the scope and status of the project. Each creates and uses its own family of data entities – its own language – with which to analyse and describe the project. These entities can be based on established hierarchical classification systems such as the Common Arrangement of Work Sections (CAWS), the various Standard Methods of Measurement (SMM), or on more ad hoc methods such as Work Breakdown Structures (WBS) and Organisation Breakdown Structures (OBS). They can also of course be based on nothing more formal that the user's personal judgement.

The problem is that all of these methods of analysis and classification operate at a minimum of two levels of detail higher than the fundamental building component level. In order to get to the CAWS or SMM level, the individual user must aggregate and massage the underlying component level data to get it into the correct shape for his or her particular purposes at CAWS/SMM/WBS level. The result, to repeat, is that the data used ends up meaning whatever the user wants it to mean. The idea of a shared language, of uniformity and consistency

Table 8.1 Information needed for construction project management

Production information	Commercial information
Scope definition	Quantity take-off
Programmes	Estimating
Method statements	Procurement
Logistics plans	Planning
Progress assessment	Cost management
Production analysis	Change control
Forecasting and reporting	Progress assessment
	Sub-contractor and supplier account management
	Commercial analysis and reporting

of meaning across the disciplines of project management, is stymied from the beginning. Home-made applications, spreadsheets, and baseless but impressive-looking planning graphics proliferate – all presenting mutually contradictory views of the project.

In terms of business systems, the most important consequence of this way of working is that, although it should be extremely desirable to do so, it is actually more or less impossible to develop applications that might unify these disciplines. This is true not just at the overall industry level or even at the level of the individual firm; it applies at the level of the individual user on the individual project. Every user applies his or her own individual judgement, talks in his or her own personal language, and everyone is talking slightly at cross-purposes. The result is that this crucial area of construction industry activity is almost entirely lacking in effective business systems; massive duplication, fractured views, unchecked – effectively uncheckable – inconsistent models of all sorts proliferate.

This is the case between the client team and main contractor, between disciplines within the main contractor's team, and also between the main contractor and the specialist trade contractors. It is almost impossible for any two people to be entirely sure that they fully and accurately understand each other. This is true at the job winning stage; it is even more the case during project execution.

8.6.2 Project delivery

The job winning process is complete; now assume the company wins the contract. The project team mobilises. That is to say, team members gradually start to come together as their commitments on current projects ease up. The acquisition team have done their job; all of the relevant documentation and full responsibility for the project are handed over to the delivery team.

At that point the project is effectively cut loose from the head office. The project people will typically move through a series of temporary facilities until the site enablement work has been completed, at which point their working home for the next two years or so will be established on site. Apart from the occasional performance review or training course, few members of the project team will set foot in the head office until the project is complete. Their world is the job site; their colleagues and friends are the members of the extended project team, including consultants, contractors, specialist suppliers and many others.

The social dynamics of a typical project team (is there such a thing?) are complex and subtle. Despite the legends to the contrary, the attitudes of construction team members are almost always inherently positive. The sorts of class-based hierarchical relationships described by Higgin and Jessop[53] are becoming a thing of the past. As in contemporary society at large, status and respect are no longer endowed simply by the order of things; they have to be earned. In a modern project team, every team member – across the entire spectrum of firms involved – is aware that his or her success is heavily dependent on the performance of others and that, to a great extent, his or her success is dependent on the success of those others. People know that even under the most basic, traditional, lump sum, fixed price contract, they must collaborate to succeed. They know that this may be difficult and challenging in many ways. But they also know that there are very few things in life more fulfilling, more satisfying and more fun than being part of a successful construction project.

So, the project gets settled in; the experienced hands get their personal standard procedures up and running, and brief and oversee the novices, as they all adjust to the routines of life on site. The team faces many particularly critical and demanding challenges during the first weeks of its existence. There are the obvious ones to do with establishing the new organisation, catching up with the details of the on-going design, physical site issues, logistics arrangements and so on. The key challenge, although it is rarely addressed explicitly in discussion of projects, is the creation of a complex, dynamic organisation, which must become capable of managing safely very high rates of expenditure, almost immediately after its inception. A large project today can reach a spending rate of several millions of pounds a month within two or three months of its start date. Only a tiny proportion of all businesses ever reach that rate of expenditure, yet construction projects do so almost as a matter of routine.

8.6.3 *The project/head office dichotomy in construction*

In most contracting companies particularly, a significant degree of isolation exists between the project and head office. In one sense this is both necessary and inevitable. The authority needed to fulfil their responsibilities must be delegated to the project's managers. Significant operational interference from head office undermines this principle and leads to confusion and delay, so most companies take a deliberately hands-off approach, unless something goes wrong.

---

53 Higgin, G. and Jessop, N., *Communications in the Building Industry*. London: Tavistock Publications, 1965; reprinted London: Routledge, 2001, pp. 92–111.

The project team may start out with standard company procedures, templates and forms and so on, but to a great extent they will make up the project accounting and reporting rules themselves. The team will usually be required to carry out its book-keeping and other basic accounting functions using a corporate system. But decisions as to which operations-level computer systems to use are often left up to the people on the project. In the commercial area, the allocation of sums in the cost plan, and allowances for items like provisions, reserves and contingencies will all be done by the project team.

The preparation of applications for payment, tender management, administration of subcontractor accounts, accounting for retentions and discounts will also typically be left to the team. Divisional or head office commercial executives may review these arrangements at the beginning of the job and periodically during the course of the work. But they generally stay at arm's length, unless something goes wrong. Other project functions, such as planning and scheduling, health and safety, and environmental management will generally be handled in a similar manner. In practice, it has to be this way.

The most important form of formal project monitoring is through weekly and monthly reporting routines. These typically address technical issues, programme status, cost reconciliation, procurement status, claims analysis, various cost and schedule forecasts, and other matters. The necessary reports are compiled from a variety of sources including the site daily diary, the commercial manager's cost management records, head office accounts reports, package managers' weekly situation reports, the planning system and others. Unfortunately, as noted in the previous section, all of these systems and documents are based on differing underlying data and all therefore present slightly differing views of the status of the project.

Arguably the most important consequence of the discontinuity of communication between head office and the project team is the general failure of construction firms to learn from their projects. As noted elsewhere, individual project team members gain enormous personal experience from every project they work on; their companies, as potentially intelligent commercial entities, learn more or less nothing. At the end of the project the slate is wiped clean; corporate memory is purged. If the information used on projects were more accurate and systematic, and depended less on individual judgement in its creation and application, there is no doubt that methods and systems could be deployed that would overcome this problem, and would enable companies to learn from their projects by accumulating information from them in a systematic, structured, reusable fashion.

8.6.4 Competition in construction

The role of competition in the free market economy is to ensure that consumers can buy products and services at the lowest price, and to induce suppliers continuously to improve the quality of the products and services that they provide. Paul Teicholz's comparative productivity graph, Figure 8.2, shows that since the early

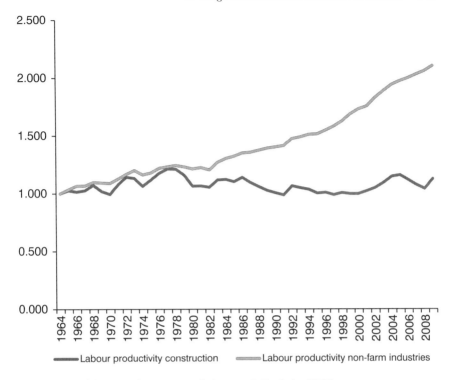

Figure 8.2 US labour productivity trends (source: P. Teicholz, CIFE)

1980s, productivity in construction in the USA has tracked significantly lower than that of all non-farm industries. He finds that, whereas industry generally has achieved a near doubling of productivity, construction output per man-hour has remained more or less constant over that period.

Professor Teicholz's calculation places the US industry in its overall economic context. The Constructing Excellence annual performance report portrays a British industry achieving little better than its US counterpart (Figure 8.3). Differences in methodology and measurement probably account for any apparent differences in performance between the two industries. Sir Michael Latham's earlier summary of a number of comparable exercises also supports these findings.[54]

These charts indicate that in the strange world of construction the rules of economic competition are somehow suspended. Competition in construction fails; it neither provides low prices for customers nor does it induce performance improvement in the industry. In other industries, firms that do not respond to the pressures of competition go out of business. In construction however, it seems that competition on price or performance is not a survival matter. So competition cannot be depended on to induce construction firms to behave innovatively.

---

54 Latham, M., *Constructing the Team*. London: HMSO, 1994, p. 63.

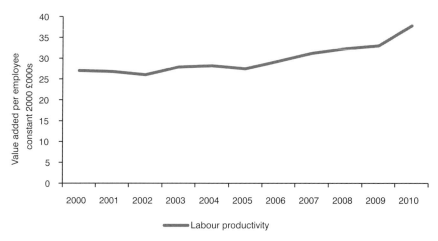

Figure 8.3 UK construction productivity (source: Constructing Excellence)

For a market to be truly competitive, it must be possible for the customer to obtain from the competing suppliers, bids which can be compared on a true, accurate and detailed, like-for-like basis. Unfortunately, as discussed in previous sections, the scope documentation which forms the basis on which construction contractors compete for work does not allow for this sort of rigorous evaluation of the contractors' proposals. The result is that contractors know they cannot be held to their bids. They know that if they come under pressure they can always find additional work or poorly documented scope on which to hang claims for additional payments. This is not a matter of incompetence on the part of the design team or even particularly a matter of dishonourable behaviour on the part of the contractors. It is simply a fact of life, entirely due to the inherently poor, untrustworthy nature of drawing-based design documentation.

Whatever one takes as being the cause, the upshot is that construction contractors do not experience competition in the same way that normal firms do. Neither of the two main operational areas of activity of construction firms, job winning and project delivery, is subject to effective economic competition. In job winning, all that's required is a plausible bid and a steady nerve. If things go awry in the project delivery phase, all that's required is a good claims team and a willingness to scrap. There is no competitive pressure on contractors to deliver projects economically, or to improve their methods of production. But equally, competent contractors have no means of securing their profitability or of defending their markets against newcomers or particularly risk-tolerant competitors. The cost of entry to the modern construction market is almost zero. Anyone can set up as a contractor and will succeed provided he or she is willing to take risks and to fight for claims; no capital is required and no sophisticated skills. This sort of predatory, risk-taking, confrontational behaviour drags everyone down until all firms in the industry compete for little more than subsistence.

8.6.5 Summary: technology and construction

The experience of other industries suggests that most business people are conservative and careful, reluctant to innovate or take risks with new technologies. Yet when the need arises they do innovate and they do take risks. The impetus to innovate can come from structural changes in their industries, from changes in the regulations they work under, and from other sources in their competitive environments. Critically, to be effective, the innovation must help firms to perform better, must give them a competitive edge of some sort in the struggle for survival in their markets.

Generally, firms have deployed information technology innovatively in three stages: initially they automated or computerised individual production operations; next they linked functionally adjacent systems, and finally they networked related systems across the span of the organisation. Each of these steps is taken in response to some particular local threat or opportunity. They are all stages in what might be described as a process of tactical diffusion of information technology. When this process is complete, the individual firm has a reasonably comprehensive information infrastructure in place; its data sets are well defined and their relationships are well mapped. This provides the platform and sets firms up in a state of readiness to embark on a second, strategic exploitation phase in their adoption of information technologies.

Crucially too, by about this stage, senior management will have become comfortable with IT. Many executives will have experience of delivering IT projects and they will be familiar with the idea of treating information as a strategic corporate asset. They, rather than the technology professionals, are usually the people who look out over the walls of their firms in search of ways of leveraging their information in partnership with neighbours in their supply chains. For these people, the flows of information between their firms and their customers and suppliers is as important as the flows of materials or services. Many of them sit on the strategic boards of their companies.[55] It is they who have used the management of information to transform their organisations and their industries.

Historically, none of this has applied to construction. There is only weak competitive pressure on construction contractors that might induce them to innovate or to take risks with technology. Because of the companies' extreme dependence on the individual judgement of their employees, project team members have to be allowed a great deal of freedom in decision making. If the team doesn't like something proposed by head office, they won't do it, or won't do it diligently. It is therefore extremely difficult for any head office function to introduce novel requirements or methods to unwilling project teams. This means that it can be extremely difficult for the firm as a whole to develop or implement significant and genuinely useful computer systems for use on their projects. The local preferences of individual project teams and team members tend strongly to

---

55 Lundberg, A., 'Wal-Mart: IT Inside the World's Biggest Company', *CIO Magazine*, 1 July 2002.http://www.cio.com/article/31174/Wal_Mart_IT_Inside_the_World_s_Biggest_Company (retrieved 10 October 2010).

overwhelm any competing corporate initiatives, however strategically important they might be.

So, to summarise: amongst construction contractors, there is no effective competition-based impetus to innovate. And even if useful innovations become available, the people at the operational front end of the industry are particularly difficult to influence, so diffusion will be patchy and erratic.

Fortunately, BIM does not depend on contractors alone for its impetus. The abolition of standard fee scales, as resolved eventually in the Enterprise Act 2002, changed the business environment significantly for architects and consulting engineers. Two things seem to have happened as a result of competitive fee tendering.

First, the consultants' scope of work is losing its traditional clarity and comprehensiveness. The professions are necessarily doing less for their now-reduced fees. In the case of MEP consultants, for example it is becoming normal to provide schematic, single line diagrams instead of dimensioned, coordinated services drawings. Architects no longer coordinate the designs of other disciplines. Consultants generally are relying more on the 'design intent' principle, and thus on the design capabilities of equipment suppliers and specialist contractors. Fully dimensioned, coordinated designs are a thing of the past. Whether this is in the interest of the client, of society, or of the professions themselves is at least debatable.

However, in this newly competitive environment, consulting firms seem to be becoming more commercial, less patrician, in their approach to business. Unlike the contractors, who as described above can slide around their contractual commitments and thus evade competitive pressure, the consultants have to grapple with competition head on. A certain amount of 'flexibility' is to be expected in interpretation of their scope of work, but this is far less, and far less invidious, than that which obtains in the contracting sphere. Crucially it does not generally engender a 'claims attitude' amongst consultants – they have only the core scope of work as a source of profit. So, paradoxically perhaps, it is the design professions, rather than the ostensibly more commercial contractors, who are being compelled by the pressure of competition to become as efficient and as productive as they can.

As the case studies in Chapter 7 demonstrate, BIM even in its most basic, stand-alone form, is one of the most cost-effective options currently at their disposal in this regard. Stand-alone models however, are just part of the tactical diffusion phase in the implementation of the BIM approach. Stand-alone BIM will probably not result in sustained profitability or defensible market share. Everyone will be able to do it; so everyone will have to do it.

However, a BIM model which is shared amongst the members of a project team is quite a different thing. Parametric component-based 3D models generate highly accurate and complete information. When these are linked and shared amongst the design team, in a BIM environment using proven standards and protocols, the whole game changes, as the next chapter demonstrates.

9 Looking forward: building with perfect information

9.0 Introduction

This chapter asks two main questions: what would the construction industry look like, how would it behave, working with trustworthy information? And, how might that situation be brought about? There are a few supplementary issues to consider, but the discussion focuses on these two. First though, it may be useful to recap briefly on the story to date.

The construction industry is performing poorly against a wide range of benchmarks. Almost all other industries have improved markedly in most aspects of their performance over the past few decades. Construction has shown no such improvement. On the contrary, in many important respects the industry is actually falling further and further behind these other sectors.

A great deal of effort has been expended by people in, and close to, the industry in trying to uncover the reasons why construction behaves as it does and in trying to correct those behaviours. This work has been going on for at least the past 60 years, but no substantial or sustained improvement has been achieved.

Buildings are amongst the most complex things produced in the modern economy, and the organisations required to design and construct them are amongst the most complicated forms of human organisation. This situation demands the very highest quality information to begin with, and then powerful, subtle information management processes to govern the communication, interchange and use of that information throughout the project team. Construction has neither of these:

- Drawing-based design, the basis of all construction information, of its nature generates very poor quality, unintelligent, un-computable, fundamentally untrustworthy information.
- Conventional contracts, organisation structures, and management procedures provide inadequate methods and frameworks for managing and sharing information effectively.

The Building Information Modelling approach promises to change all that, by providing very high quality information on which to base the design and construction of buildings and by enabling that information to be shared confidently and effectively throughout the project team. This is building with trustworthy information.

BIM tools enable designers to create comprehensive models of buildings, by inserting digital representations of real-world building components into virtual 3D space in a computer. The key characteristics of the components of BIM models are:

- They are precisely accurate in terms of their geometry: dimensions, orientation, and insertion point location and so on. They can also carry many more attributes than geometry; all of these other attributes can be equally accurate and well specified.
- BIM components are said to be 'intelligent' in the sense that they can be programmed to embody various forms of technical knowledge, including rules, such as building regulations, design standards, and others.
- BIM components are interoperable; at least in theory, objects created in one BIM system can be fully understood and reused by any other standards-compliant system.

A profound transformation in the work of most other sectors of the economy has taken place over the past 50 years or so. Manufacturing and retail are perhaps the most prominent examples. This transformation has resulted, more than any other factor, from the diffusion of computerisation and the dramatic improvement in information quality and communications capability that ensued. BIM promises to transform construction in much the same way and to much the same degree.

In other industries, IT-based transformation progressed in two broad stages:

- a tactical diffusion phase in which firms pursued internal technical efficiencies, driven largely by their finance and IT functions (phases 1 to 3 in Section 8.3);
- a strategic exploitation stage in which higher level, supply chain optimisation was the goal; this stage led to the transformation of industries and was typically led by operational business people (phase 4 in Section 8.3).

The tendency in trying to forecast how technological innovations will evolve in the future has generally been to overplay the short-term and to underplay the long-term effects. That fits with the two-stages idea. The transformative innovations discussed in Chapter 8 almost all followed this pattern. The tactical diffusion stage included the early phases during which the underlying technology became generally disseminated throughout the target industry. In these initial phases, firms operating under conditions of intense competitive pressure, invested reluctantly and cautiously, usually in order to fix a particular immediate problem: high inventory levels, high labour costs, ballooning paper handling, and such like. There was hardly ever any longer term, strategic impetus behind their decision making; short-term profit and survival were the key drivers. The ambition of IT vendors and the enthusiasm of early adopters combined to create a buzz of promotional hype and user excitement surrounding and following the release of new hardware and software products. This helped to move adoption forward incrementally, very much as Rogers' diffusion of innovations theory implies. But it did not affect things strategically.

Sometimes the new system is a new application; sometimes it's a network service. But gradually all these one-off, tactical investments add up. A point arrives when everyone has a PC on his or her desk; everyone has access to the firm's specialist piece of software; everyone has word processing, spreadsheets and presentation software; everyone has access to the shared servers; everyone has e-mail and calendars, and of course, everyone has internet access.

Having made such an investment, the company now has a real information infrastructure in place. As people in the business explore their new capabilities and learn of developments in other firms, they start to push outwards, to explore the interfaces between themselves and their commercial neighbours. Supply chains, business-to-business initiatives, e-commerce and a whole range of new business relationships, all become possible. This is the point where strategic exploitation takes off. It's a long way from the initial tactical initiatives, but would not be possible if those earlier steps had not been taken.

The transformative effects of IT innovations came about in the strategic exploitation stage. This stage sets in after the immediate problem is solved, the dust has settled, the urgent need to innovate is removed and the fundamental information technology and associated processes are reasonably well established throughout the industry or sector. By this point, an industry-wide technological infrastructure of sorts is in place, firms have mastered the details of their internal data sets and systems and they know how to build links between them. Crucially, at this stage business managers become aware of the idea of 'information' as a discrete, valuable and manageable resource. They become alert to its potential and become personally involved in the search for strategic advantage through improved information management. This stage takes on a genuinely strategic aspect, where technology people step back slightly and business managers take direct control of firms' technology innovations and investments.

The construction industry is currently somewhere in the early stages of the tactical diffusion phase with Building Information Modelling. Although there is a tendency, in this book and elsewhere, to disparage computer-aided drafting as a poor form of CAD which suppressed uptake of the real thing – computer-aided design – the construction industry has actually benefited enormously from the success of products like AutoCad. These systems, as noted in Chapter 5, being cheap and easy to use, brought CAD to the masses in just the same way that Microsoft Windows brought computing to the masses. For business people, desperate to get to grips with and take advantage of serious computing, Microsoft has been an enormously good thing. Without Microsoft, industry generally would still be in the dark ages of computing: in thrall to mainframe ogres and prey to the squabbling tribes of Unix.

The basic idea of designing with a computer rather than with a pencil is a critically important psychological hurdle to overcome in the diffusion of real CAD and, ultimately of BIM. More than any other product, AutoCad made that possible for a whole generation of designers of all sorts. Autodesk did for computer-aided design what Microsoft did for general computing. Between them, Microsoft and Autodesk products have provided the platform from which the next phase – the strategic exploitation of Building Information Modelling – can be launched.

With the exception of a small number of innovators who are investing heavily, design firms are probing the use of BIM tools, but only tentatively so. At this stage, even the innovators are using these tools primarily as advanced drafting systems. For now, the fundamental thing that the systems must do is to help designers to keep on generating drawings as efficiently as possible. The 3D modelling capability of BIM tools helps with that, as does the relative ease with which drawing files can be exchanged between disciplines and between firms. However, it is essential that firms see the more advanced information management capabilities of BIM systems as being additional to the production capabilities of their existing systems, and that no trade-off or compromise of those basic capabilities is involved in the adoption of BIM tools. Firms are concerned mainly to verify that the systems actually fit with their ways of working, and that they really do bring the short-term benefits the vendors claim. So the basic challenge in the short term is to ensure that BIM, as a form of computer-aided drafting, really works.

Things look reasonably positive. The McGraw-Hill survey data and case studies reported in Chapter 7 suggest that, despite the current economic circumstances, firms are increasingly taking the necessary steps to migrate to BIM: upgrading hardware, buying the systems, training their people and developing new procedures as required. It seems reasonable to expect that most of the medium-sized and larger design firms will have made the transition to BIM as a form of CAD within the next five years or so. That is the crucial first step in the tactical diffusion phase of this new technology.

The other players in construction – general contractors, specialists, suppliers, even quantity surveyors – all have parts to play. Apart from a few pioneers, so far there is very little evidence that these firms have become aware of BIM or its potential influence on their businesses. Contractors don't see BIM as providing any significant short-term benefit, or competitive edge, so they are not investing seriously. Some specialist firms, such as structural steel fabricators, are using 3D modelling in the same sort of way as the designers, but in general, McGraw-Hill and the case studies do not indicate any significant deployment of the information management capabilities of BIM systems.

For the next four or five years things in the world of building design will continue pretty much as they are, at least on the surface. Design firms will continue their slow, tentative migration from their existing 2D drawing-based systems to the new 3D model-based tools. The deliverables from BIM design processes will continue to be drawings and other conventional forms of documentation, although designers will gradually become aware of the information management capabilities of BIM systems. For the near future, project models will generally be developed on a stand-alone, discipline-by-discipline basis, normally with the architectural model acting as the master reference model. Coordination checking will be carried out using file import and merge routines.

Most design firms have developed standards and protocols for the use of conventional, drawing-based CAD on their projects. These are generally private documents, designed to ensure that the individual firm's internal drawing production processes and associated procedures are as efficient as possible. As the use of BIM intensifies across the industry, firms will become increasingly aware of the benefits to be gained through efficient interchange of design information, in BIM form, with their project partners. This will lead to the generalised implementation of project-wide standards – BIM execution plans – based on but broader in scope than single-practice standards, for the management of project information. BIM execution plans will preferably be based on public guidelines, such as BS1192:2007, the AEC(UK) BIM Standard, the protocol documents itemised in Section 6.3.3, or similar reference documents. Project-specific versions of these documents will generally be needed, to suit the requirements of the particular combination of firms involved in the project. The key requirement is that issues like data formats, naming conventions and presentational standards are resolved on a project-wide basis, to maximise the productivity of the overall team rather than the individual firm. The aim, as the Ryder Architects case study illustrates, will be to make it very easy, in a sense, to create virtual multi-disciplinary firms.

It may be frustrating for the vendors and BIM crusaders, but in truth this picture is not at all unhealthy. As Chapter 8 describes, the initial adoption of new technologies in other industries has always started with easily justified, tactical responses to short-term problems or opportunities. These responses diffuse and consolidate and gradually become the foundations on which longer term, more strategic exploitation can be based. So, relax and just let it happen? Well, possibly. But it would seem to make sense, rather than just drifting aimlessly towards some distant state, for the industry to take advantage of the self-improvement thinking that has gone on for over a decade now, and in that context to explore how a more considered outcome might be brought about.

The essential feature of BIM-based design is that it generates information that is trustworthy – no need to check it before using it; computable – it can be passed directly from one computer to another without human intervention; and intelligent – it can embody human knowledge and rules. For most practical purposes this is perfect information. The material outlined in Chapter 7 suggests strongly that design firms in the industry will generally be in a position to create this sort of information within the next five years or so.

At that point, the key question for construction becomes: 'What happens when you can build construction projects using perfect design information?'

9.1 Future construction

To the extent that everyone's role and experience in the industry is unique to him or her and shapes his or her expectations, so everyone in construction will have a personal view of what an ideal construction industry might look like, how it might behave and how that ideal industry might come about. The following describes one such view. It is presented in the form of a loosely specified scenario. This is not scenario planning, as practised for example by Shell.[1] As used here, the scenario approach simply provides a shaping framework on the basis of which to consider some of the key industry impacts of BIM and the related tools and techniques. The scenario as used here is not so much a forecast, more a way of thinking about what the future might bring and how the key players in construction might respond to the coming of BIM.[2]

In keeping with the recommendation of Roger Martin of the University of Toronto, a leading corporate strategist, the scenario proposes a single strategic objective for construction companies: maximisation of customer value. As Professor Martin puts it: 'Peter Drucker had it right when he said that the primary purpose of a business is to acquire and keep customers.'[3]

This scenario proposes that what the construction industry customer values and wants most is the highest quality building he or she can afford, delivered on time and within the agreed budget, carrying an unconditional ten-year guarantee, at demonstrably the lowest competitive price. In other words, he or she wants something like a modern Hyundai, Kia or Mitsubishi motor car.

This would clearly not be possible with today's construction industry. But then, neither would it have been possible with cars of the 1980s or 1990s. It has taken the automotive industry several decades of experience with a combination of innovative management techniques, enabled critically by CAD/CAM and supply chain systems, to arrive at the point where they can make this offer. Similarly, in this future BIM-based industry, buildings will be procured following a selective tendering process, on a fixed price, lump sum basis, to include for all aspects of construction and on-going maintenance, guaranteed for a stipulated period following handover and commissioning.

The building guarantee will extend to all of the features of the building and all aspects of its performance that can be tested and proven in a BIM model. This will include the construction cost and completion date, the soundness of the building fabric, its embodied carbon, thermal behaviour, energy usage, fire and smoke safety, space utilisation, lighting, acoustics, operation and maintenance requirements and costs, reconfiguration capability and costs, and other features. It will be necessary for all these tests to be carried out in the model, by the design team, prior to procurement of construction contracts.

(One way in which this might be organised would be for individual clients, or alliances of clients, to work with framework-like leagues of designers and contractors. The members of each league would cross-guarantee the projects of other league members, so that if one goes out of business, or fails to perform according to the guarantee, the others will compete for his outstanding obligations

---

1 Schoemaker, P.J.H., 'Scenario Planning: A Tool for Strategic Thinking,' *Sloan Management Review.* Winter 1995, pp. 25–40.
2 Kahn, H., *Thinking About the Unthinkable.* New York: Horizon Press, 1965.
3 Martin R., 'The Age of Consumer Capitalism', *Harvard Business Review,* Jan. 2010.

and concomitant revenues. In the same way that the professional institutes do, these leagues will also regulate the conduct and practice of their members, so that firms that under-perform against agreed standards will be disciplined, relegated to a junior league or dismissed entirely from the league system.)

Using BIM tendering, contractors will have no option but to compete directly, on the basis of their ability to provide what their clients require. It seems reasonable to suggest that what the clients of construction really want is guaranteed buildings. It also seems reasonable to suggest that as soon as one firm learns how to provide such products, everyone else in that market will have to do so too. Intelligent use of the BIM approach is the key to that capability.

But, as indicated earlier, that strategic transition point is at least five years off. Before it can be achieved, the basis on which construction contracts (without guarantees initially) are awarded will change to reflect the idea of tendering on perfect design information. At a basic level, BIM shuts off the claims tap. The construction scope of work, as described using BIM models in tender and contract, will be accurate, complete and verifiable. There will be no opportunity for claims or other forms of extras. Thus BIM-based tendering will achieve two key things: first, it will force contractors to compete on the basis of their operational capabilities; and secondly, it will very quickly eliminate predatory bidders and risk junkies. This means that competent firms will be able to compete in the expectation that their competitors, like themselves, will need to make a reasonable profit from their construction activities. Margins will no longer be driven to zero, and competent firms will be able to develop capital- or competence-based barriers to entry to their markets. This is the first step in the strategic transformation of construction.

The precise way in which this process might unfold is impossible to predict, but there are general issues that will be important in a BIM-based project process of the future. The following sections consider these in turn.

9.1.1 Design and procurement

A significant break with current practice will come with the abandonment of fast-track design and construction. Fast-track working is useful to sophisticated clients who need to keep key design decisions unresolved until as late in the project process as possible. They are prepared to accept the higher cost and greater involvement on their part in design and procurement choices that this involves. Unfortunately, aspects of fast-tracking, notably the package-by-package procurement of construction work, escaped into the wild in the 1990s. This coincided with the banning of standard fee scales and led to the situation where detailed design and dimensional coordination of building components are increasingly being carried out by the trade contractors. As noted elsewhere in this book, the consequences have generally been seriously undesirable. In future, it is to be hoped at least that buildings will be designed fully in BIM systems before the procurement of equipment and construction contracts gets under way.

This will have beneficial results for all concerned. Design consultants will carry out full services, for full fees; trade contractors will be relieved of this invidious burden, their scope of responsibility will be more clearly and more appropriately defined; and the client will get a professionally designed building. It would seem sensible for consulting engineering firms to re-take this position; BIM will enable them to do so. In this context, competition amongst designers should be exclusively on the basis of previous work, reputation and creativity, not cost.

Drawings, schedules and other conventional types of documentation will continue to be used during the early years of the transition to BIM-based construction. Initially the BIM models on which they are based will be provided mainly as supplementary information. Gradually however, as contractors become comfortable with their use and capabilities, models will be adopted as the principal form of contract documentation. The model will be used as the basis of tendering and the final version of the design model will be handed over to the successful bidder.

In the early years, two broad procurement routes, largely continuing current trends, are anticipated. First, following a complete design process, largely as outlined in the previous section, contractors will be invited to bid for the main contract, which will be awarded on a lump sum, fixed price basis. Copies of the completed design model will be provided to the bidding contractors.

The alternative is for the main contract to be awarded on a design and build basis. Consultants will develop the design to the completion of the concept stage, at which point tenders for complete design and build contracts will be invited. Again, the main contract will be awarded on a lump sum, fixed price basis, in this case the fee for the scheme and detailed design work will be included in the lump sum. The design team may or may not be novated to the contract as the client requires. Incentive/reward sharing schemes such as integrated project delivery may be used.

The guaranteed buildings approach, which should emerge within about five years, will simply add the guarantee element to these basic procurement strategies.

9.1.2 Manufacturing

Most of the innovative activity in the construction industry occurs either in the design process, driven mainly by competition of ideas, or in the manufacturing sector, driven mainly by price competition in the free market. Very little that is truly innovative happens in the layers of management, bureaucracy and construction activity that occupy the space between these two. BIM will bridge that gap, in a sense recreating the direct dialogue that took place between the architect and craftsman in years gone by. This is not a lapse into nostalgia. In any era, the architect needs to know as much as possible about the materials and components he or she is planning to have installed in the building. In the past this information comprised shared knowledge which was developed in the training of the architect and the craftsman and exchanged in close dialogue between the two. In a BIM-based industry, information of this sort will, for the most part, be programmed into the parametric objects used to create BIM models.

A large proportion of the final detailed design model on any project will therefore be made up of objects retrieved from manufacturers' online product component libraries. This places the individual manufacturer in a quandary. In order to promote the use of his equipment or materials he must make his information as widely available to specifiers as possible; on the other hand, doing so risks his material being copied by his competitors. In order to protect the manufacturers' intellectual property rights, access to their libraries may be restricted. They may make basic geometry, mass, centre of gravity and interface information available on line, but detailed design may be provided only to verified specifiers. Their libraries may also incorporate digital watermarks of some kind.

There is one big difference between the information usage patterns in construction and other industries. In all of the manufacturing industries referred to earlier, a great deal of effort goes into the design of their products and production processes, as a one-off, up-front effort. The product design stage involves exchange of relatively few, large batches of complex geometry and engineering data. These exchanges are carried out using a variety of 'standard' exchange file formats and protocols. Sometimes these formats are open, in the sense that they are published and maintained by public non-proprietary standards bodies. More often proprietary formats, usually corresponding to the software preferences of the lead member of the supply chain, are used.

However, once the production planning stage is complete and the supply chain as a whole enters into production mode, a very different mode of information exchange commences. This takes the form of streams of relatively simple, standard format EDI messages, carrying commercial, production and logistics data between the partners in the supply chain.

The information, including manufacturers' data, exchanged between members of the building design team prior to tender, corresponds with the information used in the design of products and production processes in conventional manufacturing industry. As described above, the BIM standards and protocols will govern these interchanges.

Construction will also have to develop data standards and interchange protocols to deal with the supply chain information flows between contractors and suppliers. Suppliers will be required to play a leading part in the development and implementation of construction-specific versions of the EDI messages and protocols used in mainstream manufacturing.

Given their guarantee obligations, main contractors will seek to ensure that the supply chains, linking themselves and component and equipment manufacturers, are as short and responsive as possible. Excessive sub-contracting and sub-sub-contracting will tend to die away. Manufacturers, together with their approved installers, will work closely with the main contractors, using BIM visualisation and schedule linking to develop highly accurate construction simulations and logistics programmes. Obviously the manufacturers will be required to provide guaranties for their products and will be required to guarantee the work of their installers, so only manufacturer-approved firms will be permitted to do installation work.

The UK construction market alone is unlikely to be large enough to support a thriving, innovative component manufacturing industry. Much will be standard, manufactured to store, but much will also be project-specific, short-run batches, manufactured to order. To achieve the sort of scale that makes this sort of near-mass customisation feasible, manufacturers will have to push their products into larger European and global markets. Suppliers like United Technologies, ThyssenKrupp, Schneider and Permasteelisa already sell their highly engineered products internationally. It is to be expected that the engineering content of other types of components will increase significantly in the coming years. This will happen both because of the need to control the carbon and energy embodied in building components, and also because methods of designing such components and of simulating their behaviour will improve dramatically. Materials and components like engineered masonry and glass, sophisticated *precast* concrete and its successor materials, will all be traded internationally, making economies of scale achievable.

9.1.3 Construction

The construction site of the future will be an assembly place; no manufacturing will happen there. Imagine how an industrial version of the German Huf Haus, or the Skanska/Ikea BokLok concept might work. There will be no cutting, shaping, pouring, drilling, routing, welding or grinding of material, because none of these processes can be carried out with sufficient precision in site conditions. There will be no wet trades, for the same reasons. *In-situ* concrete will be replaced completely by standard section, prefabricated beams, columns and floor and wall panels, all of which will be designed to be dropped into place by crane. Cone/ring type locators will be used to ensure accurate positioning of components and reversible locking mechanisms will be used to enable them to be demounted and reconfigured.

It was reported, as if it were a remarkable achievement, that the bearing surfaces of the prefabricated steel nodes used in the structure of the Swiss Re building were milled to a tolerance of 0.1 mm.[4] In future construction this sort of precision in the interfaces between components will be commonplace. The traditional idea of construction tolerances – plus or minus a few mm or sixteenths of an inch – will no longer apply. Components will have nominal dimensions and their connection points will be manufactured to those dimensions precisely.

The biggest problem, as ever, will be getting out of the ground. But even here, advances in surveying, soil mechanics, civil engineering and location keeping should make it possible to create foundation platforms with bearing points sufficiently precisely located to support the sort of geometrically accurate structural frames required.

Precisely formed manufactured components and assemblies of components will be shipped to site, on a strict just-in-time basis, handled once and dropped into position. Connector mechanisms will be developed that enable most of the components of building systems to be joined using click-fit, zero insertion force

---

4 Powell, K., *30 St. Mary Axe: A Tower for London*, London: Merrell, 2006. p. 84.

(ZIF) techniques, as used in the assembly of computers and other electronic devices. Systems components, like partitioning, raised floors and suspended ceilings, will be designed to be easily demountable and reconfigurable.

As much work as possible will be done under-hand. The existing pattern of off-site pre-assembly of toilet and kitchen pods and of services modules for installation in risers, ceilings and floors, will continue to intensify. Today's methods of off-site prefabrication, essentially the work of the construction site taken indoors, will tend, like other industries, to move towards the use of automated techniques, production lines and robotics as much as possible. Craft-based modes of working, even in factories, will die out.

Project organisations as a whole will become much smaller and more focused on construction operations. Working with accurate, trustworthy information will eliminate the need for large numbers of inspectors and checkers at all levels in the supply chain. The ludicrous pattern of 'man-to-man marking' in every conceivable audit and inspection role should disappear.

The reason why all this will happen in a BIM-based industry but could not happen today, is because with BIM, for the first time, contractors will have a survival-level incentive to make it happen. They will be forced to provide their products at the lowest price and will thus be compelled to compete in terms of their production capabilities. They will have no alternative but to innovate and improve their products and processes continuously.

9.1.4 Building maintenance and reconfiguration

Contractors will be required to operate their own building maintenance organisations. The guarantees they provide to their clients will lock them into the management of maintenance services on their buildings. They will not be permitted to sell off their stakes in the projects they construct. Devices such as special-purpose vehicles will not be permitted. Contractors' balance sheets will be required to account for both their construction and maintenance operations.

The aims of these arrangements are:

- To ensure that construction clients, in embarking on the greatest capital spending project most of them will experience, can call on competent, professional, profitable firms to carry out their work. But also to ensure that they can choose firms in a fair, price-competitive market
- To ensure that if one contractor fails, others can be relied upon to complete the client's project, competently, at no additional cost
- To keep the contractor committed to the building, so as to ensure he has a long-term interest in building well, with good components
- To ensure that contractors learn about how their buildings work, and can apply that learning to subsequent projects
- To ensure that contractors develop fair, collaborative, and innovative, long-term relations with their supply chains.

This way of working will also encourage construction firms to become more like manufacturers. By staying with their buildings for extended periods, contractors will make as much or more of their income from operation and maintenance services as from the initial construction work. A number of suppliers to construction, including manufacturers of switchgear, HVAC equipment, plant-rooms and lifts and escalators, already do this. The construction industry at large has much to learn from the parent companies of these firms. It should be reasonable, in this type of BIM-based industry, for contractors to brand their buildings, as other manufacturers do their products.

During the course of the project, every event in the life of any given component, from its first appearance as a conceptual entity in the earliest BIM model, through its subsequent development in detailed design, its procurement, manufacture/fabrication and installation in position in the physical building will be tracked and recorded in the BIM databases. Every planned and actual maintenance event will also be recorded in the model. The building's actual performance, aspects like energy usage, circulation patterns, acoustic behaviour, repair requirements, evaluated against the original design criteria, will also be monitored and logged continuously in the model.

A model of this sort will be a vary valuable resource. In the scenario outlined here, where the main contractor guarantees and maintains the building for an extended period after handover, the model will belong to the contractor, who will be free to add it to his 'catalogue' of projects. The industry will start using these as-built, 'catalogue' models to help new clients to visualise and test their early options. Instead of designing individual buildings or parts of buildings from scratch every time, 'catalogue' models will be used to form the basis of the detailed design of new buildings, with existing arrangements of components being reused, or selected and replaced as required to suit the preferences of the new client. Obviously the original designers of reused concepts and detailed designs will be entitled to royalty payments for the use of their intellectual property.

The 'catalogue' model will be particularly useful in that it will identify any problems which may have been encountered with the design, construction or occupation. The details of the construction process, including its budget and programme, as well as operation and maintenance costs and related issues, will also be very useful in optioneering and planning exercises.

9.2 Considerations

This section continues to use the future construction scenario asks how government, academia, the professional institutes, companies and individuals in construction should contemplate a BIM-based future industry. The section sketches out things that will have to be done by organisations and individuals. These are not recommendations, they are facts that flow from the logic of the scenario.

9.2.1 Government

Government has two important sets of dealings with the construction industry: government as client and government as regulator. In its role as client, government will take a prominent lead in demanding guaranteed buildings. Note that the requirement is not that the use of BIM should be mandatory on projects; it is that designs must be perfect and that buildings must be guaranteed. In order to achieve that, the industry will have no alternative but to use BIM and related techniques. But it will also have to incorporate elements of all of the other process improvements advocated by Latham, Egan and others: teamwork, just-in-time, lean production, build off-site, and so on.

A highly competitive, innovative, professional construction industry will emerge. The guaranteed buildings approach will remove the risk and unpleasantness of procuring buildings in the traditional way, so demand for construction will rise. The UK industry will take the lead in delivering this service in the global marketplace. It will generate higher national income in a much more transparent and taxable way. It will be a safe, high-quality industry which will employ the best young people, with the widest range of human talents, and will invest heavily in their futures. All of these social benefits, and probably more, would result from that single initiative.

In its role as regulator, the government will radically re-think the whole philosophy of regulation of the construction industry. As it stands, regulation simply adds a dead-weight cost to construction activity, in most cases with few demonstrable offsetting social or economic benefits. Contractors have no incentive to reduce this cost; they simply pass it on directly to their clients. The aim will be to arrive at a point where the industry can be as self-governing as possible, responding to appropriate, informed government intervention when required. For example, building models and actual building performance data will enable environmental regulations to be set more intelligently and more collaboratively between the industry and relevant technical experts. The contractors' on-going involvement in their buildings will align their interests with those of their customers and thus of society as a whole. This approach will help to ensure that the operational, environmental, economic and social dimensions of industry regulation will be taken more effectively into account in the drafting processes.

Like many other major client bodies, including major commercial companies, government will find itself obliged to reverse the Thatcherite policy of outsourcing its building design procurement and briefing functions.

> The public role of architecture has ... been diminished by the loss of so many local authority architecture departments and functions. The position is similar for national government. As the Urban Task Force report noted, over the last 15 years the number of qualified planners in the Department of the Environment, Transport and the Regions has fallen by 50% and architects by 95%. It also pointed out that in England, less than 10% of all architects are employed by local authorities, while the figure is 37% in Germany. No wonder

it has been so hard to integrate architectural thinking into urban regeneration in a number of places, as so much in-house expertise has been lost.[5]

This is not a case of one form of ideology versus another. It is simply a matter of fact that as the environment, and specifically the built environment, comes more to the front and centre of social and political discourse, these organisations will have no option but to have expertise in its creation and management directly involved in their strategic thinking. The management of their property assets will become a main board level issue for most major organisations. That can't be outsourced.

Government will become more seriously involved in financing and pushing forward the fundamental research needed to achieve true machine-to-machine interoperability between BIM modelling systems. For the moment, for the sort of BIM modelling that will be most commonly used over the next five years or so, data interchange between models can continue to be carried out on the basis of inspection and negotiation between the users of individual models. But for BIM to achieve its strategic potential to transform the industry, to become something more than CAD, it will be essential to complete the work on classification methodologies, data exchange standards, and interchange protocols. This work is as important as the original R&D on the fundamentals of CNC and CAD, described in Chapter 5. It requires serious commitment, not to endless international technical committees, but to something like the US grocers' Ad Hoc Committee, which set up the whole basis of EPOS, barcodes and the foundations of the supply chain industry in a period of less than three years.

9.2.2 Academia

Universities and research establishments have an important role to play and much to gain from the transformative effects of BIM. As explained in Section 9.1, the nature of the competitive forces faced by contractors will change dramatically. No longer able to survive through claims, they will have no alternative but to compete directly on their operational production capabilities. Like other mature, information-based industries, construction will then invest heavily in R&D, in its people and in its broad knowledge base. In this endeavour, it will reach out to academia for assistance in a number of areas.

First, the industry will press harder and more persistently for research and development in new building components and materials. Manufacturers will probably take the lead in much of the work on things like easy-install components, as discussed in Section 9.1.2. But it is difficult to see manufacturers of materials like steel, cement and plaster pushing hard on the development of more environmentally acceptable alternatives to their products. There will be strong demand for basic research in this and similar areas.

---

5 Worpole, K., *The Value of Architecture: Design, Economy and the Architectural Imagination.* London: Future Studies, RIBA, 2000, p. 10.

A second important area of research will be to do with the theoretical aspects of the information used in the industry. BIM-based construction will generate and capture data at the level of individual transactions, performed on individual building components, across the entire supply chain and throughout the building life-cycle. This will generate enormous quantities of data, at a very fine-grained level of detail. The industry struggles already with the problems associated with BIM data volumes and the imperative need to manage that data efficiently. The challenge will escalate in the coming years. Industry will look to academia to take a lead, to provide the theoretical underpinning for future work in the area of information analysis and classification referred to above.

For the next ten years or so, the construction industry will need all the help it can get from academia, as it first prepares for and then actually undergoes the BIM transformation. It is not likely that the historical anti-intellectual bias of the industry will be replaced overnight. But the younger, better-educated generation of managers and leaders now coming through, will need new mental models, more appropriate ways of thinking about their businesses and the industry at large. Academia must prepare for this and must take the initiative in helping firms and individuals to get to grips with these aspects of the unfolding future. Entirely new courses of study and new forms of training will be required for all areas of activity in the industry. Professionals will be multi-disciplinary, as trades will be multi-skilled. Knowledge transfer partnerships (KTPs) and other forms of interaction between industry and academia will be greatly intensified.

The industry will actually be an extraordinary laboratory for researchers across a very wide spectrum of academic disciplines, ranging from sociology and economics through business studies, engineering and materials science. But academics must get out there, into the industry, and must take part in the process. To continue to direct their thinking and writing at each other, rather than the wider industry, just won't work. As Chris Wise wrote recently: 'University research consistently fails to address the practical realities of construction … It would be so nice if some of their research was useful.'[6] A BIM-based industry will demand nothing less.

9.2.3 Professional institutions (including institutes)

It is generally accepted that the role of construction industry professionals is two-fold:[7]

- to provide expert advice to their clients, and to defend them from ill-treatment by others in the course of their projects and afterwards;
- to safeguard the public interest by ensuring that the built environment is pleasing, safe, sound and sustainable.

---

6 Wise, C., 'Academia Isn't Up to the Job', *Building*, 26 November 2010.
7 See, for example, http://en.wikipedia.org/wiki/Professional_association (retrieved 12 December 2010).

In return for carrying out these duties according to stipulated codes of professional conduct and standards of ethical behaviour, professionals are permitted legally to form exclusive-membership bodies: institutions or institutes. The principal purpose of these bodies is to create and control barriers to entry to the market for the services of their members. The professional bodies also, typically, act as learned societies, each establishing and maintaining the particular body of knowledge on which the qualifications, education and training of its members are based. So in addition to their boundaries to the outside world, the institutions establish and maintain the boundaries between professions – the discrete silos of professional practice that differentiate and separate their members from each other. They also act to represent their members to government and the general public.

In this latter role, the institutions have largely failed, as Will Hughes described it in *The Professionals' Choice*:

> The professions knew the game was up in the 1980s. The old values of public service and learned people developing their skills to the full potential were dealt a severe blow by the Restrictive Trade Practices Act of 1982, which outlawed mandatory fee scales. For the first time in living memory, professionals could undercut each other and bid competitively for work. Their clients realised that they could pit hungry professionals against each other and drive down the fees. Those who were unlucky enough to lose too many of their bids went out of business. Those who were unlucky enough to win, had to cut back the services that they offered as there was simply not enough money in the job to permit them to undertake their traditional role. At this point, the professions in the construction industry lost their grip. From that point on, they had to serve those who paid them and could no longer subscribe to the notion of public service.[8]

The long squabble with the RIBA over scale fees, initiated as far as one can see by the Monopolies and Mergers Commission in 1977,[9] was a pointless dispute over the wrong issue. It was an example of ill-informed free-market dogma swatting aside all thoughtful consideration of greater social benefit. For such an important area of the economy as the design of the built environment, the first question must be: what is the best way for the consumer to obtain the 'best' buildings and for society to obtain the 'best' neighbourhoods and cities? Logic would suggest that the answer, usually, will be for the 'best' architects and engineers to be selected according to some set of technical and aesthetic criteria, and then for them to be remunerated for their efforts according to some 'fair' basis. The selection should be made in that sequence: capability first, price very much second.

---

8 Hughes, W. 'Technological Scenario', in S Foxell (ed.) *The Professionals' Choice*. London: RIBA/CABE, 2003, p. 84.
9 Monopolies and Mergers Commission (MMC), *Architects' Services: A Report on the Supply of Architects' Services with Reference to Scale Fees*. London: HMSO, 1977.

Instead, government effectively allowed the whole matter of its relationship with building design and thus with the quality of the built environment to be managed through the crude machinery of competition and fair trade. Typifying this was the appointment of the legal and economics consultancy firm LEGC Ltd to carry out a review of competition in the professions, by the Office of Fair Trading. The LEGC report in 2000[10] and subsequent 2001 OFT[11] ruling spelled the end. The contest was finally 'won' by the Office of Fair Trading with the threat of litigation contained in its Progress Report of 2002.[12]

Most of the fears relating to fee competition raised by the RIBA from the beginning, have subsequently come to pass.[13] And there is no evidence of any significant counterbalancing benefit to society or the consumer of architectural services.

The polarisation in practice size between global giants and barely viable minnows that has taken place over the past 20 years has been a huge social and strategic blunder. It seems intuitively desirable that there should exist a reasonably even distribution of firm sizes amongst designers. Government and industry must search for a way to restore the traditional pattern and to enable design firms to compete and flourish in all sensible size bands. A good starting point would be for all built environment design work to be procured under a UK version of the American Brooks Act method of Qualification Based Selection in which technical proposals and fee quotes are presented and examined entirely separately from each other.

But that is about the selection and future of individual professionals and their firms. The perpetuation of the professional institutions is a different matter. The conventional rationale says that the existence of the institutions is justified on broad and important economic and social grounds. The main economic justification for the institutions is their role in protecting construction clients from the effects of the profound information asymmetries that are involved in the decision to build and all that follows. The basic argument is that clients are naïve, and unless they are accompanied and protected by insiders, they will be taken advantage of by the industry. This justification disappears with a BIM-based construction industry. The relationship between client and building provider that will obtain in a BIM-based industry will be much more like that which exists today between a customer and the provider of any relatively ordinary complex product. There are two specific points. First, with rich models, the number and nature of the decisions that the customer will have to make will be much more within the capability of the lay client than is the case today with drawing-based designs. Second, the nature of the commercial relationship between the customer and building provider will be a much cleaner one, with full scope definition, no claims, no extras. Thus, even

---

10 LEGC, *Restrictions on Competition in the Provision of Professional Services: A Report for the Office of Fair Trading*, London: LECG, 1977.
11 Office of Fair Trading, *Competition in Professions: Report by the Director General of Fair Trading (OFT328)*. London: Stationery Office, 2001.
12 Office of Fair Trading, *Competition in Professions: Progress Statement (OFT385)*. London: Stationery Office, 2002, Para. 3.9, p. 11.
13 MMC, Para 139, p. 49 and elsewhere in Chapter 7.

without guaranteed buildings, the need for 'professional' guidance and protection will be greatly reduced.

The main social justification for the institutions is the public-interest one. However, in a BIM-based industry, the public interest in the safety, soundness and sustainability of the built environment will be accommodated in software. Modelling systems will use (or specify the use of) only safe, sound and sustainably acceptable products and will enable complete building designs to be checked for compliance with all relevant codes and regulations before they go to contract.

So the two fundamental justifications for the privileged roles of the professional institutions become redundant. The idea of the professionals as being separate from, more trusted, somehow ethically superior to their industry colleagues, derives largely from the poor quality – the untrustworthy nature – of the information on which the operations of the traditional industry are based. In a BIM-based industry, when information becomes accurate, true and trustworthy, the need for elitist custodians of the client's and wider society's interests disappears.

More specifically, the need for the institutes and institutions disappears. Many of the classic responsibilities of professionalism will persist. The general obligation to create harmonious places, to provide beautiful, sound buildings that respect the environment, and to do so in a way that also respects the environment and the industry's customers, will continue to depend mainly on the professionals. But in a sense, in a mature BIM-based industry, almost everyone in construction will be a 'professional' of some sort. In this situation the two concepts: 'professional' and 'institution' become separated and the question of whether the things now called institutions have a continuing role becomes moot.

No doubt some of the institutions will survive in their roles as learned societies. But ideally as many have argued for many years, there will be a single mother body that will embrace professionals from all of the areas of design, manufacture and construction that make up the industry. The most valued institutions will be those that work to create networks of networks that support and stimulate their members and deliver services that they really need. The most useful and important of these will be expertise and leadership in helping their members to develop the interfaces and to negotiate the protocols that will enable them to connect to each other quickly and seamlessly to form the sort of 'virtual multi-disciplinary firm' that Ryder Architects describe in the case studies in Chapter 7. This is a very different mode of operation, but is how most design firms will operate in the BIM-based industry of the next decade. The institutions will evolve to reflect that fact. Or not.

9.2.4 Firms and projects

Building Futures, a joint initiative between the Royal Institute of British Architects (RIBA) and the Commission for Architecture and the Built Environment (CABE) facilitated the development of two studies: *The Professionals' Choice*[14] in 2003 and

---

14 Foxell, S. (ed.), *The Professionals' Choice: The Future of the Built Environment Professions*, London, CABE/RIBA, 2003.

'Practice Futures'[15] in 2010. Both of these used scenario techniques to sketch out a variety of visions of the future, ostensibly dealing with the construction professions generally, in fact focusing on architecture. A striking feature of the futures painted in all of the scenarios in these studies is their bleakness; not black pessimism perhaps, but visions dominated by a sense of threat, of futures to be feared and even resisted: '… an era of rampant shareholder power …' in which '… the profession would become little more than an anti-capitalist pressure group'.[16]

The future of building design firms will not be like that. As Chris Wise describes it:

> From now on, creative people should have a great time. Before building anything we will be able to frolic in a virtual world: there is no risk in ideas … if a virtual idea falls down, we can learn from it and try something else. Ideas will be tested in a series of 'what-if' scenarios on a palmtop, and soon on an ear-mounted brain sensor, then in a holographic force-field. The arrival of interactive design software has revolutionised the way we design things … it means that engineering has become more of an art, architecture more of a science, and all design more intuitive. This crossover is at the heart of some of the most innovative design thinking today.[17]

This is the sort of experience that 'creative people' can look forward to. Quite contrary to most of the sentiment of the Building Futures scenarios, the technology will not oppress them or allow their firms to do so. BIM will liberate people and firms from the constant stress of abusive contractual liability and obscure commercial threat. The key to its being able to do that is that it will liberate everyone, not just the creatives, but also the management people and production people. Anyone doing a job that cannot be done by a machine or a computer will be liberated and uplifted. In construction – particularly on projects – most useful jobs are like that.

A pair of important and, even now, very interesting reports: *Building Britain 2001*,[18] and *Investing in Building 2001*,[19] were published in 1988 and 1989 respectively, by the Centre for Strategic Studies in Construction of Reading University. Significantly, some of the most senior people in UK construction were actively involved in much of the actual work on this project. The first report used three scenarios to project the industry forward about ten years, to 2001. The envisaged scenarios were briefly as follows:

- An 'Expert Client Scenario', essentially business as usual, in which the industry continues in its traditional, fragmented way, investing little,

---

15 http://www.buildingfutures.org.uk/projects/building-futures/practice-futures (retrieved 4 November 2010.)
16 Davies, W. and Knell, J., in Foxell, pp. 136–7.
17 Wise, C. in Foxell, p. 36.
18 Bennett, J. and Flanagan, R., *Building Britain 2001*. Reading: Centre for Strategic Studies in Construction, University of Reading, 1988.
19 Bennett, J., Croome, D. and Atkin, B., *Investing in Building 2001*. Reading: Centre for Strategic Studies in Construction, University of Reading, 1989.

responding flexibly at a subsistence level of profitability, to relatively expert, exploitative clients. This was the scenario that the writers expected to be the most likely outcome – depressingly true.
- Second was a 'Consortium Scenario' in which, recognising that small and medium-sized firms are a good thing, temporary alliances and consortia were envisaged as being the principal form of project organisation, similar in effect to partnering and PFI arrangements.
- Third was the 'Big Building Firm Scenario' in which a few large companies – contractor-based – dominate the industry, giving it coherence and political substance. Still missing today.

All three scenarios stressed the client demand for single-point responsibility. The first achieved this by having the client taking very direct control over its own construction, as was the case with firms like Canary Wharf, BAA, Stanhope and some of the shopping centre developers. A second feature was the demand for fully guaranteed building life-cycle services – a little like PFI. Information technology was a vaguely described but significant issue in all three scenarios.

It is interesting to note that the consortium scenario was reported on by Marco Goldschmied (Richard Rogers & Partners) and David Bucknall (Bucknall Austin & Partners) and the big builder scenario was championed by Brian Hill (Higgs & Hill plc) and Martin Laing (John Laing plc).

Why did neither of the forward-looking visions of the future come to fruition? There are two main reasons. First, because it would have been extremely difficult to do, both technically and commercially. Secondly however, and as with all similar visions for construction, because even if it had been possible, no one who could do anything about it had any incentive to do so. The construction industry is not a sentient, self-aware, responsive entity, capable of thoughtful analysis or rational behaviour. It's more like an amoeba, groping through its surroundings, responding autonomously to the stimulus of threats and opportunities in its environment. Exhortation is neither stimulus nor threat.

The three Reading scenarios remain quite valid. Future construction will include elements of all three. However, BIM changes the ground rules sufficiently to kick the second and third ideas back into life. It does so by providing 'perfect information' for bidding construction work and by making it possible subsequently for firms – contractors or consortia – to provide guaranteed buildings.

Firms working in and with the construction industry will see a two-stage process unfold. As outlined above, first there will be a gradual consolidation of the technologies and operational protocols. This will happen over the next five years or so and will be driven mainly by the designers. During this period, these firms will have an opportunity to take control of the construction industry, as the consortium scenario implies. However, the period will not last for long. As soon as the capability is generally in place in the industry, 'perfect information' bidding will dramatically energise the contracting side of the industry. Contractors, with greater commercial experience, broader knowledge of industry supply chains and easier access to capital than the designers, will bring the Big Builder scenario to fruition. It would be a good thing if manufacturers took part in this, but on past form this is unlikely.

These are macro-scale developments, changing things in the overall industry. At the micro level, the level of the individual organisation, firms of all types – designers, contractors, installers, fabricators, manufacturers – will pursue two strategies. First, in recognising that construction really is an 'information industry', they will start to prioritise their management of information as their principal corporate asset. Information managers, not IT managers or CAD managers but serious business people, will be on or very close to most boards of directors, as retailers and other sectors have shown. For firms in construction, as opposed to less information intensive industries, it will be particularly important that the information used in the business is well understood, well specified and well managed.

Firms will recognise increasingly that large proportions of their total information resource originates outside the walls of the corporation, with customers, suppliers and collaborators. This will lead to the second strategic development, which will require that firms look outwards more systematically and more strategically. The key will be to create the sorts of organisational and technical interfaces that will enable them to engage with their industry neighbours in a 'plug-and-play' fashion, connecting and disconnecting quickly and painlessly as the organisational designs of the projects on which they work require.[20]

9.2.5 Individuals

Machines leverage muscle; computers leverage calculation; BIM leverages imagination. Machines can complement muscles or they can substitute for them. In most of industrial manufacturing the substitution effect had predominated; human muscle power is largely redundant. In a similar way, computers can both complement and substitute for human calculation skills. And in a similar way to machines and muscles, computers now broadly substitute for human calculation, particularly for structured, programmatic forms of calculation.

BIM modelling tools can be used to complement imagination, by enabling the designer to test ideas easily and quickly and to communicate them to other people accurately. But, at least for the foreseeable future, there is little likelihood that BIM – or anything else for that matter – will ever be able to substitute for imagination. As Albert Einstein said: 'Imagination is more important even than knowledge. For knowledge is limited to all we now know and understand, while imagination embraces the entire world, and all there ever will be to know and understand.'[21] So for the foreseeable future, all sorts of people of imagination,

---

20 For an authoritative treatment of the design of projects, as job-specific organisations, see Morris, P.W.G., *The Management of Projects*. London: Thomas Telford, 1997, p. 213ff.
21 Originally in an article 'What Life Means to Einstein', *Saturday Evening Post*, October 26, 1929; reprinted in Einstein, A., *Cosmic Religion: With Other Opinions and Aphorisms*. New York: Covici-Friede, 1931, p. 97, quoted in Calaprice, A. (ed.), *The Expanded Quotable Einstein*. Princeton, NJ: Princeton University Press, 2000.

including but not limited to creative designers, will have great, BIM-leveraged fun, as Chris Wise suggests.

There are other human faculties, such as judgement (incorporating attributes like instinct, ethics and intellect) and persuasion (intuition, empathy, communication) whose owners can take advantage of BIM techniques without any threat of their being substituted by BIM systems. These faculties are required in a wide range of activities in construction. All three are required for sound decision making, of which, as outlined in Section 8.2, a great deal goes on in this industry. All three are also required in negotiation and doing deals, again a great deal of which goes on in construction. So people with these skills will prosper. They can take full advantage of BIM in their work without any fear of being replaced by such systems. Others, lacking un-programmable talents, will not be so fortunate.

The previous section pointed out that successful firms will be those that can look out beyond their corporate boundaries and readily form production alliances with others in their supply chains. In a similar sense, successful individuals will be those who make the effort to learn how people in adjacent professions to their own work and think, and who can build links with their neighbours, quickly and easily. These are not so much multi-disciplinary people, but rather people who understand their own information requirements well, and also understand what other people will want to do with the information that they provide.

If it ever really existed in construction, the era of the focused individual working in isolation from all around him in the project team is over. As Mario Guttman, vice-president of HOK, put it in relation to design teams: 'I don't think we can continue to include pure CAD drafters or CAD-illiterate design professionals. I think we will all be better design professionals as a result.'[22] A plethora of specialist tools will be released in the coming years, enabling everyone in construction to work with BIM models in ways that suit their particular project roles, so everyone will have to become skilled in the use of these BIM-based tools. CAD is just a tool, BIM is just another tool, but by connecting them together – people and their tools – something greater results.

9.3 The global picture

The twenty-first century will be the century of the city. The proportion of the population living in towns and cities in Europe has been increasing slowly and erratically since before Roman times. A more dramatic and sustained process of urbanisation is generally regarded as having started with the Industrial Revolution in Britain, in the second half of the eighteenth century. Lewis Mumford's description of early industrial towns and cities is a bleak one:

---

22 http://www.aecbytes.com/viewpoint/2005/issue_17.html (retrieved August 4, 2010).

... a blasted, de-natured man-heap adapted, not to the needs of life, but to the mythic 'struggle for existence'; an environment whose very deterioration bore witness to the ruthlessness and intensity of that struggle. There was no room for planning in the layout of these towns. Chaos does not have to be planned.[23]

Today, a third of the people who live in cities live in slums[24] little different to the places Mumford describes; one person in every six of the entire human population is a slum dweller. Over the coming 50 years, the total human population will continue to grow towards its forecast maximum of about ten billion people, and an increasing proportion of the total will be city dwellers. The UN calculates that, sometime in 2008, a cross-over took place, as shown in Figure 9.1. From that point onwards the numbers of people living in cities outnumbers those in the countryside. By 2050, roughly 70 per cent of the population of the planet will live in cities. Given that almost all of the growth of cities in the next half-century will be in developing countries, the proportion of the human race living in urban poverty can be expected to increase dramatically.

For some people it will come as a surprise to learn that this is regarded as an almost unambiguously good thing by most of the experts involved in human development. A recent UN report on the state of the world population demonstrates this convincingly, for most of what one might call the key performance indicators of human development. 'Cities concentrate poverty, but they also represent the best hope of escaping it ... The potential benefits of cities far outweigh the disadvantages: The challenge is in learning how to exploit its possibilities.'[25] Radical new theories of urbanism are emerging in response to this challenge.[26] There is also an interesting on-going debate about the urban poor, the 'Bottom of the Pyramid' regarded by C.K. Prahalad as 'resilient and creative entrepreneurs and value-conscious consumers', but as potential victims of an excessive reliance on markets by his critic Aneel Karnani.[27]

However that particular argument plays out, the general point is that views of the role of the city in human development are changing dramatically – for

---

23 Mumford, L., *The City in History*. New York: MJF Books, 1961.
24 Population Division of the Department of Economic and Social Affairs of the United Nations Secretariat, *World Population Prospects: The 2006 Revision and World Urbanization Prospects: The 2007 Revision*. http://esa.un.org/unup (retrieved 10 November 2010), p. 16.
25 UNFPA. *Unleashing the Potential of Urban Growth*. New York: United Nations Population Fund, 2007. p.1.
26 See for example: Bettencourt, L.M.A., Lobo, J., Helbing, D., Kühnert, C. and West, G.B. 'Growth, Innovation, Scaling, and the Pace of Life in Cities'. *Proceedings of The National Academy of Sciences of the USA*, 2007, 104(17). http://www.pnas.org/content/104/17/7301.full.pdf+html (retrieved 10 November 2010).
27 Prahalad, C.K., *Fortune at the Bottom of the Pyramid: Eradicating Poverty through Profits*. Philadelphia, PA: Wharton School Publishing, 2005, as quoted in: Karnani, A., *The Bottom of the Pyramid Strategy for Reducing Poverty: A Failed Promise*. New York: UN Department for Social and Economic Affairs, 2009.

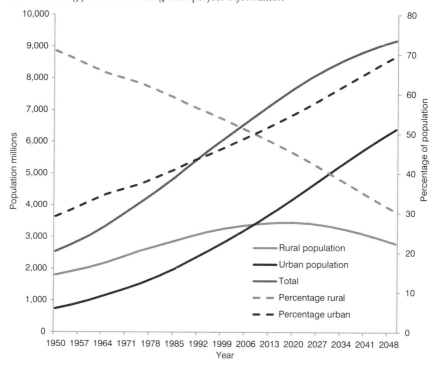

Figure 9.1 World population projections (source: UN World Population Prospects, 2007)

the better. A powerful summary of the argument is provided in Stewart Brand's surprising book *Whole Earth Discipline*,[28] which captures the sense of logical tension that the process represents in today's context. Brand and other eco-pragmatists are convinced that if the great problems of ignorance, poverty, injustice, and inequality are to be solved, the city is where it will happen. Observing the horrors of slum life, urban sprawl and environmental degradation that accompanies the growth of cities, it seems remarkably prescient or astonishingly optimistic that towards the end of his great book, Mumford was still able to assert that: 'The city's active role in future is to bring to the highest pitch of development the variety and individuality of regions, cultures, personalities.'[29]

The basic economic mechanism that operates here is that, by bringing people together into large conurbations, urbanisation provides everyone with more opportunities to trade – whatever they've got – with other people. Karnani insists that the state has a significant role to play in poverty reduction and, more importantly, in the development of 'legal, regulatory, and social mechanisms to protect the poor who are vulnerable as consumers'. The UN Population Fund report too, emphasises the need for institutions to protect the underprivileged,

---

28 Brand, S. (2009).*Whole Earth Discipline: An Ecopragmatist Manifesto*. Viking Press, New York.
29 Mumford, p. 570.

with a particularly interesting reference to the evils of: 'Speculators (who) hold on to land in and around the city, expecting land values to increase. They do not bother renting, especially if they fear that users might gain some rights to continued use, or controlled rents.'[30] Clearly another situation where a site value tax system might be beneficial.

This huge process, this picture of the future of humanity, represents the greatest of all the tests facing the global construction industry. The first part of the challenge is how to create or dramatically expand the capacity of hundreds of cities, largely in the developing world, in such a way as to create beautiful places where the mass of people can live rewarding and productive lives. (A total construction programme equalling about a thousand times the size of metropolitan Birmingham will need to be completed in the next 40 or 50 years.)

The second part of the test for construction is to do all this without devastating the planet. Recall that even in the relatively environmentally aware UK:

- Buildings are responsible for almost half of the country's carbon emissions, half of our water consumption, about one third of landfill waste and one quarter of all raw materials used in the economy.[31]
- Between them, mining, quarrying, construction and demolition account for 62 per cent of all the waste generated in the UK.[32]
- The UK Office of National Statistics, Environmental Accounts 2010, breaks down the British economy into ten major sectors. Construction is the only one of these to have increased its greenhouse gas emissions per unit of output, between the base year 1990 and 2008.[33]

This is obviously not an encouraging starting point. To take the single issue of climate change, it has been established for some time that global warming and atmospheric carbon levels are correlated. The chart in Figure 9.2 is a simplified version of one presented in a *Nature* magazine article by J.R. Petit and others in 1999.[34] Some debate persists about the direction of causality, but the fact of the relationship is generally agreed.

A key argument of the majority of the climate science community is that the increase in global warming recorded since 1880, as shown in Figure 9.3, coincides with the increase in industrial activity in Western economies over that period.

---

30 UNFPA., p.48.
31 *Strategy for Sustainable Construction*, 2008, UK Government, Department for Business, Enterprise & Regulatory Reform, Construction Sector Unit, p. 4.
32 http://www.massbalance.org/resource/massbalance p. 8.
33 http://www.statistics.gov.uk/downloads/theme_environment/EnvironmentalAccounts2010.pdf Table 1. Retrieved 10 November 2010.
34 Original data at: J.R. Petit, J. Jouzel, D. Raynaud, N.I. Barkov, J.-M. Barnola, I. Basile, M. Bender, J. Chappellaz, M. Davisk, G. Delaygue, M. Delmotte, V.M. Kotlyakov, M. Legrand, V.Y. Lipenkov, C. Lorius, L. Pe´pin, C. Ritz, E. Saltzmank & M. Stievenard. Climate and Atmospheric History of the Past 420,000 Years From the Vostok Ice Core, Antarctica. *Nature* Vol. 399, June 1999.

202 *Looking forward: building with perfect information*

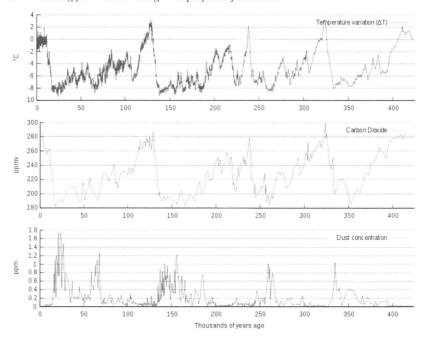

Figure 9.2 Vostock ice core data (temperature, CO_2, dust content) (source: Wikipedia)

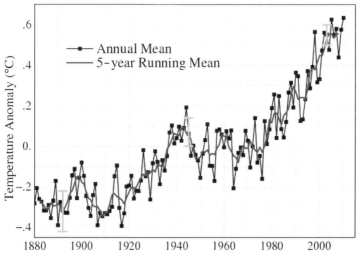

Figure 9.3 Global warming observations 1880–2010 (source: GISS/NASA)

The main characteristic of industrial activity that connects with global warming is the production of greenhouse gases (GHC), notably carbon dioxide, CO_2. The inference drawn by most experts in the field is that continued production of GHCs, at anything like historical rates, will have catastrophic effects on the environment:

> We conclude that ... the planet as a whole, is ... within 1°C of the maximum temperature of the past million years. We conclude that global warming of more than 1°C, relative to 2000, will constitute 'dangerous' climate change as judged from likely effects on sea level and extermination of species.[35]

So to restate the challenge: the construction industry must build more cities and supporting and connecting infrastructure in the coming 50 years than has been built in all of previous human history, for an increasingly demanding, well-informed public, and must do so without wrecking the planet. That's the challenge. Of course, it's also the opportunity.

The larger UK firms – architects, consultants and contractors – are well known and trade comfortably in the global construction marketplace. And there has also been a certain amount of foreign involvement in the UK market in recent years, again primarily large firms pursuing large projects. Despite the EU's surprising lack of success in bringing about significant rationalisation or unification of the European construction market, it is certain that the construction industry will become increasingly global in the coming decades. Increasing world trade regulation, coupled with increased use of BIM-type technologies and concomitant increase in contract bid and award transparency, will flatten playing fields around the world. Increasing use of component-based BIM in the design of buildings will bring increasing requirement for the manufacture of components in factories in the real world – anywhere in the real world where economies of scale can be achieved. Design methods, component standards, manufacturing capabilities, and construction methods will converge globally. Britain's historical links, the English language, the industry's international experience and reputation, and its relatively early adoption of BIM, combine to give UK construction a significant advantage in this version of the globalisation process.

Recent achievements of UK architects, engineers and contractors in creating remarkable buildings in China, the Middle East and elsewhere, demonstrate the ability of British firms to excel in the production of huge, single-building, trophy projects, using BIM-like methods. And true mega-projects will come; developments like Arup's proposed Dongtan eco-city suburb of Shanghai, where firms – whether they be Consortia or Big Builders, or something else entirely – will be required to create complete neighbourhoods, towns and entire cities. But the most important developments in the global marketplace will come when all construction, everywhere, is carried out using the sort of 'perfect information' that the BIM approach makes possible. It is almost impossible to imagine how much of a contribution to the future of mankind and the planet such a form of construction might make.

---

35 http://pubs.giss.nasa.gov/docs/2006/2006_Hansen_etal_1.pdf Retrieved 10 November 2010.

Figure 9.4 Earthrise (source: NASA)

Index

2D drawing systems 66
2D objects 59
3D modelling 59, 65, 69, 78, 81
3DS Limited 132
3D Studio Max 78
30 St Mary Axe. *See* Swiss Re building

AceCad 89
ACIS 68–69, 69
Acropolis CAD 71
activity planning models 39–40
ADAM 63
Ad Hoc Committee on a Uniform Grocery Product Code 152–154
AEC. *See* architecture, engineering and construction (AEC)
AECOM 123
Aish, Robert 71, 77, 78
American Institute of Architects (AIA) 92, 107
American Institute of Steel Construction (AISC) 100
American National Standards Institute (ANSI) 148, 153
AMR Research 144
analogue building 77
Andrew Scott Ltd 132
Apollo 69
Apple Macintosh 70, 74, 75
applications programming interfaces 99
Applicon 69
A&P Supermarkets 152
APT programming language 63
Archicad 74, 89, 129
archi-tecton 22, 87, 118
architects; BIM adoption 111; changing role of 22–23, 86
architectural design; traditional 86
architectural drawings; clients' understanding of 35; consistency of 31; production of 30–31
architecture, engineering and construction (AEC) 66; CAD 71–72, 79, 83, 101; design 60–61; industry 70
Arup 76, 77, 203
attributes of parametric components 44–45
AutoCad 70–71, 71–72, 99, 159, 179, 180
Autodesk 60, 70, 75, 78, 79, 87, 89, 99, 100, 115, 180; case study 120–124; DXF format 101
Autodesk Seek 78, 86, 95
Automatically Programmed Tool (APT) NC programming language 63
automation of manufacturing 141
Autor, David 161, 162, 163
Auto-trol Technology 69
Aveva 70

BAA 196
Barbour Index 20, 78, 86
Barcelona Fish 76, 117, 118
barriers to entry 25, 183
Beck Construction 72
Bentley Systems 70, 75, 89, 115, 126; case study 124–127
Bézier, Pierre 67, 68, 69, 79
bid conditioning 38
bidding process 37–38, 95, 167
Big Building Firm Scenario 196
bills of quantities 36–38, 42, 57
BIM; 3D modelling 180; adoption 109, 113–115, 163; benefits 180; and bidding 95; and collaborative working 93; communication with clients 47–48; communication with design team 48–49; and construction management

50–54; contract procurement 50; data formats 98–99; definition of 81–82; demonstration of 6–7; design production and administration 46–47; dissemination of information 45; and embedded knowledge 86–87; in Europe 113–115; and imagination 197–198; implementation 91–92, 106–107; improvement in information quality 26–27, 28; information flows 101–103; information management 91–106; information quality 8; and intellectual propery 105; lump sum basis projects 94; and management forms of contract 94; modelling systems 3–4; parametric objects 72–73; and project functional areas 81; and project management 52; quality of information 88, 165, 177–178; reference model 82; and standards 4–6; standards and protocols 185; system characteristics 82–88; tendering 183; three layer approach 83; trustworthy information 11; in USA 109–114
BIM authoring tools 82–90, 107–108
BIM-based construction; transition to 184
BIM-based design; advantages 46–49; certainty of outcomes 48; general features 44–45
BIM execution plans 181
BIM model 50; and build owner/occupier 54; characteristics 178; computability 51, 57; consistency of 48; key benefits 45; role in constructions operations 51–52; single unitary model 81; software tools 58
Bird, Bryn 124
Bird's Nest, Beijing 3
Bison Manufacturing. 133
'blobby' architecture 76, 77
Bluewater retail centre 92
boat building 67
Boeing 6, 66, 99
boundary representation (BREP) 68
Bovis Construction 92
Braid, Ian 68, 79
Brand, Stewart 200
Bristol Myers 152
Bristol University 71
British Aircraft Corporation 66
Brown, Stephen 150, 151, 152
Brynjolfsson, E. 142
Bucknall Austin & Partners 196
Bucknall, David 196

budget overrun 23, 24, 39
BUILD-1 68
Building Design Partnership 71
Building Electrical Systems 89
Building Futures 194, 195
Building Magazine 15, 24
building maintenance 187–188
Building Mechanical Systems (CADDUCT, DDS) 89
building owner/occupiers; advantages of BIM to 10, 54
building reconfiguration 187–188
BuildingSMART 5, 100
Buro Happold 76
business process re-engineering (BPR) 142

CAD 58–80, 61, 179; advantages 65; and design administration 34; applications 61; challenges 59–62; computer requirements 79; credibility of drawings 34; drawing process 64; elements of 65; file exchange 35; history 62–80, 69, 71–72; industry size 69; industry structure 69; and personal computers 70–71; systems 1, 147, 169
CADAM (computer-graphics augmented design and manufacturing) 69
CAD/CAM 51, 61–62, 145, 159, 182; data 149; processes 69; software 63
CADCentre 70
CADD 61; data exchange 61; deliverables 61
Calma 69
CAM. *See* computer-aided manufacturing
Cambridge University 66, 68
Canary Wharf 196
Carmarthenshire County Council 132
CATIA (Computer-Aided Three-dimensional Interactive Application) 69, 75, 76, 88, 117, 118, 159
central reference model 91
Centre for Strategic Studies in Construction 195
Chandler, A.D. 137
CICC 26
CIMsteel 100
CIS/2 (CIMsteel Integration Standard Version 2) 100
cities 198–200
Citroën 66, 67
clash detection 51
clients; advantages of BIM to 8–9, 47–48; attitude to construction industry 24; communication with 35, 47

clipping algorithm 65
closed-loop control systems 149
CNC 159; data 62; machines 61, 63; machine tools 8; manufacturing 4
collaborative working 17–18, 76, 92–96, 103, 120–121, 182; forms of contract 28
collateralised debt obligations (CDOs) 156
Commission for Architecture and the Built Environment (CABE) 194
Common Arrangement of Work Sections (CAWS) 169
communication; with clients 35, 47–48; with contractors 35; with design team 35, 48–49
communications channels 139
competition in construction 36, 172–174
competitive fee tendering 176
component 3–4; BIM model 50; industry foundation classes (IFCs) 5; libraries 105–106; parametric 73–74, 83; parametric properties 74; properties of 84–86; schedule 50
component-based model 71
computability 51
Computer-Aided Acquisition and Logistics Support (CALS) 99
computer-aided design. *See* CAD
computer-aided design/computer-aided manufacturing. *See* CAD/CAM
computer-aided drafting and design. *See* CADD
computer-aided manufacturing 62–63
computer integrated manufacturing (CIM) 2, 10, 51, 62
computerised numeric control. *See* CNC
Computervision 69
Consortium Scenario 196
Constructing Excellence (CE) 12, 17, 23, 27, 173
construction; as an assembly process 53; as an information intensive industry 12
construction contracts, 2008 15
construction industry; and academia 190–191; competition in 50, 172–174; European consolidation 203; future prospects 181–187; future scenarios 195–197; and government 189–190; information technology in 166–176; production management 54–57; productivity 173; project/head office dichotomy 171–172; quality of information 1; regulation of 189; underperformance 81, 136, 177

Construction Industry Council (CIC) 93
construction management 94; and BIM 50–54; and information quality 39–43
Construction Management Association of America (CMAA) 109
construction output 19
Construction Project Information Committee (CPIC) 106
construction project software 88–90
construction project software map 90
constructive solid geometry (CSG) 68
contractors; advantages of BIM to 9–10; communication with 35, 49; main 92–94; project risk 25; specialist 94–97; UK construction 14–15
contract procurement 50
Contracts in Use survey 20, 93
Coons, Steve 99
coordinated reference models 105
coordinate system 64
Cortada, James 137–138, 140, 141, 156, 158, 160
cost management 172
cost overruns 52
cost planning 167
cost planning techniques 42
CPM-based project management 54
CPM networks 39
craft-based working 2, 22, 62–63, 187
craft component of on-site operations 22
Crane, Mike 124
creative destruction 165
Crowley, Andrew 100
customer relationship management (CRM) 167

Dassault Systèmes 66, 69, 75, 76, 117
data entry devices 64
data exchange 81, 91, 97–101, 148–149; proprietary standards 6; standards 4
data formats 79, 98
Data General (DG) 69
data-integrated supply chains 6
data-interchange standards 2
Davenport, Thomas 142
David Chipperfield Associates (DCA) 126
Davidson, John 71
Day, Martyn 79
DBSim 129
de Casteljau, Paul 67, 68, 79
decision making; significance of 164–165
design administration; advantage of BIM 46–47; CAD systems 34; problems 34
design and build 20, 93–94, 184

DesignBuilder, 129
design change control 4
Design Data 89
design intent 48, 60, 63, 72, 75, 86, 95, 176; architect's 61
design processes 59–62; information flow 32–35
design production; advantage of BIM 46–47; problems 34
design teams; comunication with 35, 48–49
Diehl Graphsoft 75
diffusion of innovation 138, 179
Digital Equipment Corporation (DEC) 69
Digital Hand 137–138
Digital Project 76, 88, 89
digital revolution 137–138, 158
Dongtan eco-city 203
DOS operating system 70
drafting systems 58–59
drawing-based design 30–32, 44, 177; checking of documents 2, 32, 34; levels of information in 32; limitations of 1–3, 7–8, 34, 45–46; processes 43; quality of information 169
drawings; in AEC design 60–61; consistency 31; dumb 61; interpretation of 36; misinterpretation. 31; process 64; production 58–59; synchronisaton of 59; with computer 63–66
Drucker, Peter 161, 182
dysfunctional competition 38

Eastman, Chuck et al. 82, 87
Edge Structures 132
Egan Report 17–18, 27, 28
Egan, Sir John 27, 189
Einstein, Albert 197
electronic funds transfer systems (EFTS) 156
electronic point of sale (EPOS) 6, 10–11, 29, 51, 150, 152, 159
engineering analysis 61
engineering manufacturing 147–149
Enterprise Act 2002 176
enterprise resource planning (ERP) 76, 144–145
Escher, M.C. 66
Expert Client Scenario 195
external networking 145–146
extranet 92

Fairmont Foods 152
family-based contracting firms 19

Family Mosaic Housing Association 128
fast-tracking 183–184
file formats 81, 91
financial services sector 155–157
flexible manufacturing systems (FMS) 147
FMI Research Survey 109–110
Ford 6
format sharing 79
Forrest, Robin 68
Fortran 68
Foster & Partners 3, 76, 77
Gantt charts 39
gbXML 129

GDS 71
Gehry Associates 3
Gehry, Frank 76, 88; case study 116–119
Gehry Technologies 76, 89, 115
General Electric 99
General Foods 152
general manufacturing 146–147
General Mills 152
Generative Components 77, 78, 87
Gherkin, The. *See* Swiss Re building
Gintran 71
global warming 201–203
Glymph, Jim 76, 88, 116, 117, 118
GM (General Motors) 66
GMW 71
Goldschmied, Marco 196
graphical user interface 65
Graphisoft 74, 89, 113, 115; case study 128–131
Great Belt Fixed Link 124
Green BIM report 112–113
Green, Robert 114
Grimsby University Centre 123, 124
guaranteed buildings 182, 184, 187, 188, 196
Guggenheim Museum, Bilbao 3, 76, 118, 119
Guttman, Mario 198

Haley, Mike 79
Hamil, Stephen 114
Hanratty, Patrick 63
Harvard University 66
Hemsley Orrell Partnership 128
Hepworth Gallery, Wakefield 126, 127
Herzog & de Meuron 3
Hewes Associates 24
Higgin, G. and Jessop, N. 28, 171
Higgs & Hill plc 196
Hill, Brian 196

H.J. Heinz 152
HOK Architects 46
Huf Haus, 186
Hughes, Will 192

IFC 5, 6, 100; format 86; standards 100
Ikea 6–7
impossible object 66
Industrial Revolution 18
industry foundation classes. *See* IFC
information flows 96, 97, 101–103, 148, 185
information interchange protocols 91, 101–103
information management 141–142; extranet 92; strategy 92
information quality 26, 28, 52, 57, 136; and bills of quantities 37; in BIM models 51; and construction management 39–43; and design processes 32–35; and procurement process 36–38
information technology; adoption in industry 140–146; in construction industry 166–176, 175–176; in engineering manufacturing 147–149; in financial services sector 155–157; in general manufacturing 146–147; in processes industries 149–150; in retail 150–154; relative cost 160; social consequences 160–166; tactical diffusion of 175; and transport industry 158
information technology sector 157
Ingram, Jonathan 71
Initial Graphical Exchange Specification (IGES) 99, 149
innovation; attributes of 139; diffusion of 138–140, 139, 141; project-led 15
Innovaya Composer 89
integrated project delivery (IPD) 17–18, 92–93
intellectual property 103, 105–106
intelligent components 3, 44, 74, 178
Interact 70
Interactive Graphics Design System (IGDS). 70
inter-bank transfers of cheques 155–156
Intergraph 69, 70, 75
internal integration 143–145
International Alliance for Interoperability (IAI) 100. *See also* BuildingSMART
International Standards Organisation (ISO) 100

interoperability 79, 91, 97–101, 106, 122, 178, 190

Jacquard loom 63
job winning 167–170, 174
John Robertson Architects (JRA) 128
just-in-time 2, 147, 189

Kalay, Y.E. 75
Karnani, Aneel 199, 200
key performance indicators (KPIs) 12, 27
Kidder, Tracy 70
Kier Eastern 124
KLH Massivholz 125
knowledge; as a state of mind 164; implicit and embedded 86–88
knowledge transfer partnerships (KTPs) 191
Kochan, D. 63
Kolarevic, B 75, 88
Koons, Steve 68
Kroger Supermarkets 152

Laing Construction 25
Laing, Martin 196
Laing O'Rourke Northern 126
Lang, Charles 68, 79
Latham Report 26, 27, 28
Latham, Sir Michael 173, 189
layered contracts 37
lean production 2, 147, 189
learning from projects 41–42
Leeds University 100
Lipman, Robert R. 100
Liverpool University 71
Llanelli Scarlets Rugby Stadium 132–134
local optimisation 142–143
Lockheed Corporation 66, 69
lofting 67, 117
logic diagrams 59
London Stock Exchange 20
Lubetkin, Berthold 120
lump sum basis projects 36, 94, 184
lump sum forms of contract 20
machine tools 62

MacLeamy, P. 46–47; MacLeamy curve 46
management contracting 20, 94
Manchester Central Library 123
Manufacturing and Consulting Services (MCS) 63
manufacturing resource planning (MRP II) 143
Martin, Roger 182

mass customisation 2, 186
material scheduling systems 61
materials requirements planning (MRP) 143
Max Fordham 128
maximisation of customer value 182
McDonnell-Douglas 66
McGraw-Hill Construction 78
McGraw-Hill surveys 110–113, 180
means and methods statements 60
mechanical CAD (M-CAD) 71
mechanical design 59–60
MEP engineers; BIM adoption 111–112
Mercedes-Benz 66
method statements 60
Michaels G. et al. 162
MicroStation 70, 77, 132
Miller Partnership 132
MiniCad 75
MIT 63, 64, 66, 68, 99
model-based design 6, 30, 83; advantage of 7–10
model-based information 8
modelling; solid 66–68; surface 66–68; wireframe 66–68
modelling standards; 5
modelling systems 58–59
model ownership 103, 105
Monopolies and Mergers Commission 192
Morandi, Giorgio 116
Morledge, R. et al. 23, 93
Morris, Peter 39, 168
Morton, R. 18; and Ross, A. 23
multi-disciplinary design review and integration cycles 48
Mumford, Lewis 198
Murray, M. and Langford, D. 16–17, 28

NASDAQ 157
National Building Specification 114
National Institute for Standards and Technology (NIST) 99
National Stadium, Beijing 3
Navisworks, 78
NC; machines 60, 68, 147; systems 60; tools 67
Nemetschek AG 75
neutral exchange format 99–101
Newell, Dick and Newell, Martin 70, 79
New York Stock Exchange (NYSE); 156
Nissan 66
Noble, David 143, 161
non-uniform rational basis splines 67
Norfolk County Council 124

normalisation exercises 38
Northrop 66
Norwich Open Academy 124, 125
numerical control. See NC
NURBS. See non-uniform rational basis splines

objects; parametric 72–74
Office of Fair Trading 193
offshoring 162
off-site manufacture 11, 125, 131, 133, 187
Olympic Village, Barcelona 117
optimism bias 41
Øresund Bridge 124
Organisation Breakdown Structures (OBS). 169
Oxford Regional Health Authority 71

PADL-1 (Part and Assembly Description Language) 68
parametric components 4, 45, 81, 184–185; behaviour of 106; embed information 88
parametric modelling 72–74, 83, 86
parametric objects 72, 72–74
Parametric Technology Corporation (PTC) 71
Parasolids 68–69
Parc y Scarlets 132–134
Parmiter Street development 128–131
Partington, Robin 77
payment flows 25
PDES (Product Data Exchange Specification) 100
Pemberton remediation site 132
Permasteelisa 118, 186
Pittman, Jon 60
PIX Project protocol 106
planned progress curve 55
planning systems 41
planning versus forecasting 40
Plant Design Management System (PDMS) 70
PLM system 75
point of sale (POS) systems 29
Prahalad, C.K. 199
predatory bidders 38, 50, 183
predictability 23–25, 26, 52, 136
prefabrication of building components 11
price competition 36
process design 59
process engineering 59
process flow diagrams 59

process industries 149–150
Procter & Gamble 154
procurement 183–184
procurement documentation; 36–38
procurement process 47, 49; and information quality 36–38
procurement strategy 91, 92
production management 54–57
product lifecycle management (PLM), 76
ProENGINEER 71
professional institutions 191–194
profitability 23–25, 26–27, 136, 174, 176; contracting margins 1987–2009 25; of core scope of work 38
profit sharing 93
project delivery 170–171, 174
project design teams, avantages of BIM to 9
project/head office dichotomy 171–172
project information flows 96, 97
project management 42–43, 52–53, 171–172; optimism bias 41; systems 41
project management activities; and information quality 39–43
project management information 170
project organisation 91, 92
project partnering 17, 93–94
project predictability 24
project process model 32, 33
properties of components 84–86
proprietary data formats 101
PRO/Reflex 72
publishing industry 148
punch tape/cards 63

quality assurance 2
quality of data 107, 168–169
quality of information 165, 169, 177

Ramboll; case study 124–128
Rappoport, A. 101
Redcar and Cleveland Council 121
Reflex 71, 72
Renault 66, 67, 69
representation of objects 64–65
retail sector 150–154
Revit 78, 80, 121, 122; Revit Architecture 89; Revit MEP 89; Revit Structures 89
Revit Technology Corporation 72
RFIs 125, 134
RIBA (Royal Institute of British Architects) 192, 193, 194
RIBA Plan of Work 103, 104, 106, 129, 131

Richard Rogers & Partners 196
Rifkin, Jeremy 161, 166
Rogers, E.M. 138, 141, 144, 179
Ross, Doug 63
Rowecord Group 132
RUCAPS 71, 80
Ryder Architecture 181; case study 120–123
Ryder, Gordon 120

Satoh, A. 18
schedule overruns 23, 24, 39, 52
schedule targets 39
schematic diagrams 59
Schumpeter, J.A. 166
SDRC 75
SDS/2 89
Securities and Exchanges Commission (SEC) 155
self-awareness of components 85
Shanghai Tower 112
Sheppard Robson 124
Siemens 75
Silicon Graphics 69
Simon, Herbert 161, 162
single purpose entities (SPE) agreements. 93
Skanska/Ikea BokLok 186
Sketchpad 64, 65
skill levels in construction 22
slums 199
Smart Geometry 76, 78
Smith, Adam 137
software; internal integration 143–145; local optimisation 141–142; stand-alone applications 141
solid modelling 68, 69, 78
Sonata/Reflex 71, 113
splines 67–69
stand-alone software applications 141–142
standard classification systems 89
Standard Method of Measurement (SMM) 37
standards-based interchange of data 30
Stanhope 196
STEP (Standard for Exchange of Product Information) 100, 149
stock-keeping units (SKUs) 151
strategic exploitation of information technology 175, 178, 179, 180, 181
strategic partnering 17, 93–94
StruCad 89
structural engineers; BIM adoption 111
Structural Modeller 89

Sun Microsystems 69
supervisory control and data acquisition (SCADA) systems. 150
supply chain management 145–146, 147, 157, 158, 198
supply chains; data-integrated 6; information flows 11
surface modelling 67–68, 69, 83
Sutherland, Ivan 64
Sweets 78
Swiss Re building 3, 77, 186

table of offsets 67
tactical diffusion of information technology 175, 176, 178, 179, 180
Taft–Hartley Act 148
take-offs 89
Team Homes Limited; case study 128–131
technical drawing. *See* drawings
Teicholz, Paul 172–173
Tekla 89, 115; case study 132–134; Tekla Structures 89, 132; Tekla X-Steel 89
Tesco 154
ThyssenKrupp, 186
Toyota 66
transport industry 158
trustworthy information 11–12, 51, 61, 154, 181

UK construction industry 14–29; analysis of 16; competition in 36; conservatism of 18; development and history 18–21, 53; dysfunctional competition 38; strategic challenges 23–24; structure 14–16; top firms 21
UN/EDIFACT. 148
unique product code 151–155
United Technologies 186
Universal Code Council (UCC) 153
universal/unique product code (UPC) 6, 150, 159
University Campus Suffolk project 113
University of Rochester, 68
University of Toronto 182

Unix 69, 70, 179
UN Population Fund 200
untrustworthy information 1–2, 11, 28
urbanisation 198–202
USA construction industry 17–18
US Air Force 158
US Department of Defense 99
US Food and Drug Administration 152
US National Bureau of Standards, 99
US National Institute of Building Sciences 107

value added networks (VANs) 149
Vectorworks 75
Victoria Hall 122
virtual 3D space 3
virtual buildings 4
Virtual Construction (VICO) 75, 89, 129
Voelcker, Herb 68

Wakefield Waterfront 126, 127
Walker, John 70
Wal-Mart 154
Watson, Alastair 100
Watts, John 71, 79
Weisberg, David 74, 76, 77, 99
Weisman Museum, Minneapolis 117
what-if scenarios 4
Whitby and Bird 124
Winch, G. 28
window drawing 65
wireframe modelling 66–67
Wise, Chris 191, 195, 198
Wolstenholme, A. 27
Work Breakdown Structures (WBS) 169
world population 200

X12 standard 148
X21 158

Y2K scare 157
Yates, Peter 120

zero insertion force (ZIF) techniques 186

Taylor & Francis
eBooks
FOR LIBRARIES

ORDER YOUR FREE 30 DAY INSTITUTIONAL TRIAL TODAY!

Over 22,000 eBook titles in the Humanities, Social Sciences, STM and Law from some of the world's leading imprints.

Choose from a range of subject packages or create your own!

- ▶ Free MARC records
- ▶ COUNTER-compliant usage statistics
- ▶ Flexible purchase and pricing options

- ▶ Off-site, anytime access via Athens or referring URL
- ▶ Print or copy pages or chapters
- ▶ Full content search
- ▶ Bookmark, highlight and annotate text
- ▶ Access to thousands of pages of quality research at the click of a button

For more information, pricing enquiries or to order a free trial, contact your local online sales team.

UK and Rest of World: online.sales@tandf.co.uk

US, Canada and Latin America:
e-reference@taylorandfrancis.com

www.ebooksubscriptions.com

Taylor & Francis Group

A flexible and dynamic resource for teaching, learning and research.